Anticipation Science

Volume 3

Editor-in-chief

Roberto Poli, University of Trento, Trento, Italy

Anticipation Science encompasses natural, formal, and social systems that intentionally or unintentionally use ideas of a future to act in the present, with a broad focus on humans, institutions, and human-designed systems. Our aim is to enhance the repertoire of resources for developing ideas of the future, and for expanding and deepening the ability to use the future. Some questions that the Series intends to address are the following: When does anticipation occur in behavior and life? Which types of anticipation can be distinguished? Which properties of our environment change the pertinence of different types of anticipation? Which structures and processes are necessary for anticipatory action? Which is the behavioral impact of anticipation? How can anticipation be modeled?

The series is interested in receiving book proposals that:

- are aimed at an academic audience of graduate level and up
- combine applied and/or theoretical and/or philosophical studies with work especially from disciplines within the human and social sciences broadly conceived.

More information about this series at http://www.springer.com/series/15713

D. John Doyle

What Does it Mean to be Human? Life, Death, Personhood and the Transhumanist Movement

Springer

D. John Doyle
Department of General Anesthesiology
Cleveland Clinic
Cleveland, OH, USA

ISSN 2522-039X ISSN 2522-0403 (electronic)
Anticipation Science
ISBN 978-3-030-06942-1 ISBN 978-3-319-94950-5 (eBook)
https://doi.org/10.1007/978-3-319-94950-5

This Springer imprint is published by the registered company Springer Nature Switzerland AG
The registered company address is: Gewerbestrasse 11, 6330 Cham, Switzerland

To my wife of four decades, Jo-Anne, and our son, Jonathan.

Foreword

It has been a pleasure and a challenge undertaking the writing of this book. My hope that the book will provide the reader with valuable new insights into the ethical debate on transhumanism and human enhancement.

The book is distinctive in that it breaks new ground in a number of domains pertinent to transhumanism, especially with respect to matters of death and dying and with respect to cryonic suspension. In addition, since the book is written by a trained medical scientist (M.D./Ph.D.), the book naturally places a special emphasis on scientific and clinical rigor, particularly with respect to citing the peer-reviewed medical literature.

That being said, one special challenge faced in the writing of this book was that it necessarily addressed two very different audiences: philosopher/ethicists on the one hand and clinician/scientists on the other.

With respect to the first audience, I have done my best to avoid unnecessary technical jargon, although I readily admit that this is at times simply unavoidable.

With the second audience in mind, I have included an unusually comprehensive review of bioethics in Chap. 2.

Finally, here is my advice to those who are fascinated by the ideas presented and wish to learn more: start by consulting some of the recent books listed in the Appendix as well as some of the review articles listed in the references provided at the end of each chapter. In addition, the transhumanist website humanityplus.org is filled with wonderful resources such as their **h+** magazine that will undoubtedly interest many readers.

D. John Doyle
Department of General
Anesthesiology
Cleveland Clinic
Cleveland, OH, USA

Acknowledgements

I owe a sincere debt of gratitude to a great many individuals who have helped shape my thinking in what has been a lengthy philosophical journey that began when I took my first philosophy course at St. Francis Xavier University (Canada) at the tender age of 16. Fifty years have passed from that time, but my interest in philosophy has only grown, thanks to the various mentors who have helped me along the way.

I am particularly grateful for helpful discussions and advice from Prof. Benda Hofmeyr and Prof. Alex Antonites, my doctoral dissertation advisors at the University of Pretoria.

I am also indebted for helpful insights from Dr. Martin Harvey and Dr. Allyson L. Robichaud of the Department of Philosophy, Cleveland State University, Cleveland, Ohio, USA, as well as from Dr. James Hughes of Trinity College, Hartford, Connecticut, USA.

I am especially grateful to my wife Jo-Anne and my son Jonathan, to whom this book is dedicated, for being supportive of my life-long academic interests.

Finally, I am indebted to the numerous individuals who advised me as the book evolved over the last decade, and particularly those who identified misstatements, errors, and omissions. Naturally, I am responsible for any remaining errors.

There are instances where we have been unable to trace or contact the copyright holders. If notified the publisher will be pleased to rectify any errors or omissions at the earliest opportunity.

Some of the material in this book is drawn from a series of journal articles I previously published in Ethics in Biology, Engineering and Medicine:

- Doyle DJ. An Introduction to Bioethics and Ethical Theory. Ethics in Biology, Engineering and Medicine. 2010; 1:19–41.
- Doyle DJ. What Does It Mean to Be Human? Humanness, Personhood and the Transhumanist Movement. Ethics in Biology, Engineering and Medicine. 2010; 1:107–131.

- Doyle DJ. Pharmacologic Emancipation from the Human Condition: Laudable Goal or Dangerous Pipe Dream? Ethics in Biology, Engineering and Medicine. 2010; 1:199–214.
- Doyle DJ. The Thorny but Pervasive Problem of Permissible Deaths. Ethics in Biology, Engineering and Medicine. 2010; 1:259–265.
- Doyle DJ. Should Logic Trump Intuition in Bioethical Discourse? Contrasting Peter Singer and Leon Kass. Ethics in Biology, Engineering and Medicine. 2011; 2:1–9.
- Doyle DJ. Life, Death and Brain Death: A Critical Examination. Ethics in Biology, Engineering and Medicine. 2011; 2:11–31.
- Doyle DJ. Cryonic Life Extension: Scientific Possibility or Stupid Pipe Dream. Ethics in Biology, Engineering and Medicine. 2012; 3:9–28.
- Doyle DJ. Robots, Androids, and Cyborgs in Warfare: Ethical and Philosophical Issues. Ethics in Biology, Engineering and Medicine. 2014; 4:13–23.
- Doyle DJ. Exploring the Intersection of Religion, Psychology, and Ethics. Ethics in Biology, Engineering and Medicine. 2016; 6:91–107.

Contents

About the Author

D. John Doyle is an Anesthesiologist at Cleveland Clinic, in Cleveland, Ohio, where he holds the rank of Professor. He is currently seconded to Cleveland Clinic Abu Dhabi, where he is a Consultant in the Department of General Anesthesia.

He received the M.D., Ph.D., and FRCPC qualifications at the University of Toronto in Toronto, Canada in 1982, 1986, and 1986, respectively. Although he has numerous scholarly publications in the fields of medicine and anesthesiology, in recent years he has developed his long-standing interest in philosophy by branching into research and academic writing in the fields of ethics and bioethics, culminating in being awarded the D.Phil. degree in Moral Philosophy from the University of Pretoria in September 2017.

About the Author

List of Figures

List of Tables

Abstract

This book addresses various aspects of the question "What does it mean to be human?", approaching the issue from biological, philosophical, and ethical perspectives. The discussion focuses primarily on the philosophical and ethical implications of the transhumanist movement, an intellectual community seeking to favorably transform the human organism via the safe deployment of interventions such as genetic engineering and pharmacologic enhancement.

Following a review of bioethical principles, the book begins by examining the notion of "personhood" in philosophical, historical, and biological contexts. Next, the possibility that developments in neuropharmacology might ultimately lead to an artificial paradise free of the negative aspects of the human condition (but without the often-destructive effects of today's mood-altering drugs) is considered. A philosophical difficulty related to the existentialist notion of "authenticity" in such a "mood optimized" synthetic existence is identified.

This discussion is followed by a rather technical exposition on the occasional difficulties of establishing when a person is alive or dead, particularly in the setting of "brain death." A number of philosophical flaws with the notion of brain death as it is used currently are presented.

The discussion then considers some of the philosophical issues raised by the possibility of human cryonic suspension. A distinction is made between the information preserved in a person's brain and the substrate used to hold that information. Implications for personhood are also discussed, as well as a number of related ethical issues.

Next, objections raised by "bioconservative" critics of transhumanism are critically examined and found to be for the most part unconvincing, frequently relying on emotion and intuition rather than on evidence, logic, and reason.

Finally, in the penultimate chapter, the relationship between religion and transhumanism is explored.

Chapter 1
Introduction

Homo sapiens, the first truly free species, is about to decommission natural selection, the force that made us.... Soon we must look deep within ourselves and decide what we wish to become.

Edward O. Wilson, Consilience, The Unity of Knowledge. (http://www.paklinks.com/gs/philosophy-and-spirituality/179723-post-brainy-quotes-4.html)

Abstract This chapter introduces the reader to the central themes introduced in the book, beginning with some historical vignettes and insights from the world of science fiction. These themes are introduced to help explain why it can at times be problematic to define what it means to be human or to establish which organisms merit full human rights or full moral standing. Also introduced are the central themes of the transhumanist movement, the notion of artificial lifeforms and the implications for the notion of personhood. Finally, the chapter introduces the concept that some forms of death are potentially reversible as well as introduces the cryonics movement that depends on this particular premise.

1.1 What Does It Mean to Be Human?

"What does it mean to be human?" This question has been a matter of considerable philosophical debate for centuries. The answer will depend in part on the "lens" through which the question is examined. For instance, a geneticist might define humanness in terms of an organism's genetic composition, looking for DNA sequences unique to humans, while a zoologist might use an organism's behavior (such as looking at the use of language or tools) as a guide to humanness. Yet another scientist (a physical anthropologist, say) might focus on bone structure. Other people, including many scientists and philosophers, might argue that, in real life, formal definitions may be less useful than the "I know it when I see it" approach common to everyday life.

D. J. Doyle, *What Does it Mean to be Human? Life, Death, Personhood and the Transhumanist Movement*, Anticipation Science 3,
https://doi.org/10.1007/978-3-319-94950-5_1

After the "discovery" of America in 1492, considerable speculation arose as to whether the natives of these lands were indeed human, and many Conquistadors and colonists justified the cruel mistreatment of American natives on the basis of the argument that they were subhuman (de las Casas 1992). To set the issue straight from a theological point of view, in 1537 Pope Paul III issued a Papal bull entitled *Sublimis Deus* which declared American Indians to be rational human beings with a soul, and thus being appropriate for conversion to the Christian faith. Thus perhaps began the debate within the intelligentsia as to when a particular living creature is entitled to be considered to be human.

Many years later a similar issue arose in the US Supreme Court in the context of the "Dred Scott Case" (Fehrenbacher 1981; Graber 2006; Maltz 2007). At issue was whether Dred Scott, a slave who had lived in states and territories where slavery was illegal, should be considered to be a free man or should instead be forcefully returned to his owner. In 1857 the Supreme Court ruled 7 to 2 against Scott. Chief Justice Roger B. Taney delivered the majority opinion, arguing that that any person descended from black Africans, whether slave or free, could not be a US citizen and was thus not entitled to the human rights ordinarily assigned to American citizens, stating that blacks were: "beings of an inferior order, and altogether unfit to associate with the white race, either in social or political relations, and so far inferior that they had no rights which the white man was bound to respect".[1]

This gravely flawed legal decision took over a century to be fully remedied, and will always exist as a regrettable stain in American legal history. In its day this case also served to stir the debate as to when a particular human individual is entitled to be considered to merit "personhood" in the eyes of the law.

In more recent decades, the treatment of the Jews in Germany and elsewhere during World War II serves to remind us that despite the enlightenment that was hoped to come with the 20th century, moral corruption and intellectual laziness can lead entire nations to hold a particular ethnic group to be subhuman and unworthy of life, even though these individuals were not always easily distinguished from the rest of the population. Consider, for instance, the following provocative and inflammatory quotations from Hitler's *Mein Kampf* [2]:

- "The personification of the devil as the symbol of all evil assumes the living shape of the Jew."
- "Was there any form of filth or profligacy, particularly in cultural life, without at least one Jew involved in it? If you cut even cautiously into such an abscess, you found, like a maggot in a rotting body, often dazzled by the sudden light—a kike!"
- "The great masses of the people will more easily fall victim to a big lie than to a small one."

Issues such as the Dred Scott case, the Papal bull *Sublimis Deus* and the treatment of German Jews in World War II raise ethical and philosophical issues about personhood and moral status that are introduced in Chap. 2 and are discussed throughout this work.

[1] http://www.digitalhistory.uh.edu/documents/documents_p2.cfm?doc=23.

[2] http://www.allgreatquotes.com/mein_kampf_quotes.shtml.

1.2 Insights from Science Fiction

For many decades, science fiction writers have written stories about sentient computers, about intelligent robots serving humans as companions, about cyborgs made from a blend of human and machine parts, and about fully organic humanoids who were not at all human in any traditional sense. For instance, Isaac Asimov's famous book of essays *I Robot* (Asimov 1991) (and many of his other writings as well) dealt with stories of "positronic" robotic brains and artificial intelligence, usually in a setting of space exploration. Asimov even proposed an interesting form of robot ethics (Sawyer 2007):

1. `A robot may not injure a human being or, through inaction, allow a human being to come to harm.`
2. `A robot must obey orders given to it by human beings, except where such orders would conflict with the First Law.`
3. `A robot must protect its own existence as long as such protection does not conflict with the First or Second Law.`

A more recent cultural reference might include the fictional android character Lieutenant Commander Data, of **Star Trek: The Next Generation** fame. In this movie series, Data is a sentient artificial life form with extraordinary cognitive abilities, although he is still unable to feel emotions or understand many human idiosyncrasies. (This was, apparently, a direct consequence of his specialized programming). Under what circumstances would entities such as Lieutenant Commander Data merit the rights and protections of ordinary human beings? Who would be offered citizenship?

1.3 Transhumanism

Transhumanism is a relatively recent intellectual and cultural movement that promotes an interdisciplinary approach to understanding and evaluating potential means for enhancing the human organism, the human race and the human condition (Bostrom 2003; Fukuyama 2002; Zehr 2015). Expressed in broad strokes, transhumanism seeks to produce "posthumans" who will variously live longer, be smarter, be stronger, be more peace-loving, or otherwise constitute an improvement over regular humanity in some nontrivial manner. It aims to achieve its goals primarily through technological innovation in fields such as genetic engineering, neuropharmacology, computer technology, artificial intelligence, and molecular nanotechnology. Some of the less mainstream transhumanist themes include efforts to develop conscious, self-aware computers, efforts to reanimate individuals deemed to be dead by ordinary clinical criteria, and even altering humans with a view to facilitating space colonization. Transhumanism is the subject of Chap. 3 in this book.

1.4 Artificial Life

Another issue discussed in this book is the quest for "artificial life". In May 2010 scientists at the J. Craig Venter Institute announced that they successfully produced a synthetic cell controlled by human-made DNA (Gibson et al. 2010; Wu and Tan 2014). Although the practical aims of this work are ostensibly to develop better ways to clean polluted water, speed up vaccine production and so on, one can imagine how the techniques involved, might, in principle, eventually allow scientists to produce humans free of undesirable genes, produce humans with multiple copies of certain desirable genes, or even produce humans with entirely new genes aimed at extending the natural life span of humankind. Conceivably, it could even lead to the development of new animal/human hybrid life-forms.

While unquestionably an important scientific breakthrough, this development raises a rich variety of bioethical issues. While this breakthrough technology might soon yield bacteria capable of cleaning toxic waste dumps, and numerous other things beneficial to humankind, the dangers should also be appreciated. No doubt, it could also be used in the future to help make the most powerful bioweapons imaginable, even (plausibly) weapons targeted at specific ethnic groups carrying ethnically unique genes.

Another feared scenario is the possibility, admittedly remote, that a laboratory accident at a research facility investigating new life forms might wipeout life on earth as we know it. This is the theme in the 2007 movie *I am Legend* where an accidental plague based on a genetically engineered virus aimed at eliminating cancer kills most of humanity and transforms the rest into ogres, while the sole survivor (played by Will Smith) struggles heroically to find a cure. These issues are also discussed in Chap. 3.

1.5 Cosmetic Neurology

Recent scientific developments in the neurosciences—especially advances in neuropsychology and neuropharmacology—are yielding exciting treatments for often crippling neurological diseases like narcolepsy, epilepsy, Alzheimer's disease, Pick's disease and Parkinson's disease. While these treatments are aimed at patients with neurological disease, it is now apparent that that many of these treatments may also be applied to "normal" individuals free of disease as a means of achieving "quality of life" improvements. This enterprise is sometimes called "cosmetic neurology" (Chatterjee 2004; Cakic 2009; Launis 2010; Frati et al. 2015). Such enhancements might variously make one more intelligent, more able to memorize large amounts of information, better at mental arithmetic or at chess, more personable, and so on. The technical, safety and bioethical issues associated with the possibilities offered by pharmacologic enhancement is the subject of Chap. 4.

1.6 Personhood Theory

In recent decades a field of philosophical study has emerged, referred to as "personhood theory" (Jones 1984; Gillon 1996; Takala 2007). Influenced in good measure by the writings of John Locke, it holds that for an entity to be a person, that entity must variously be conscious, self-aware, take purposive action, have the ability to reason, possess the ability to communicate, and/or meet other criteria, such as having a "longitudinal identity". The term "entity" is used here rather than "human" to deal with the possibility that there may exist nonhuman persons. For instance, based on the argument that many apes have consciousness, the Great Ape personhood movement is an effort to create recognition of chimpanzees, bonobos, gorillas, and orangutans (the great apes) as *bona fide* legal persons on the basis that they have mental abilities similar to humans (Cavalieri and Singer 1994). As an example, the late chimpanzee Washoe, who lived from 1965 to 2007, was the first non-human to learn to use American Sign Language for communication. Washoe even passed on some of her knowledge to her adopted son, Loulis. Similarly, the recently deceased gorilla Koko had a sign language vocabulary of over 1000 words, which she used in formulating complex statements and questions. She even has her own legacy web site: http://www.koko.org.

To many philosophers, drawing on Lockean criteria for personhood, persons may not necessarily be humans (some animals may merit personhood) while some human entities such as embryos, fetuses or severely brain damaged adult patients may not qualify for personhood. Under such a scheme, some of the great apes may qualify for personhood while some people afflicted with severe dementia from Alzheimer's disease or severely impaired cognition as a consequence of traumatic brain injury may not.

A related question is concerned with when a human embryo becomes a human being (Suckiel 2008; Chan and Harris 2010). Certainly, in the early stages of pregnancy some people might say "this is not yet a human being, only a small mass of cells", while others might say "this is a potential human being, but not yet a full human being." This question has been the subject of enormous debate (Gilbert 2004; Kamm 2005; Wood 2005; Evans 2005; Hug 2006; Suckiel 2008; Chan and Harris 2010; Uppal et al. 2012).

Yet another associated issue is "Can there be degrees of humanness?" To reflect on this question, consider that some "properties of state" are binary in nature while others permit many conditions. Binary biological properties allow only two states, such as: alive versus dead, pregnant versus nonpregnant, HIV positive versus HIV negative. Other properties of a biological system may more allow for many more, or even an infinite number, of states. Examples here include: height, weight, age, and so on. But which of these two approaches best applies to the notion of humanness? These and many other related issues are also the subject of this book.

Table 1.1 later in this chapter provides a list of thought-provoking questions related to humanness and personhood that has captured the imagination of a great many philosophers. The list is offered to illustrate the vast number of deep philosophical

Table 1.1 Some philosophical questions related to humanness and personhood, some of which are addressed in this book

• Does being "human" have the same meaning as being a "person"? Are all persons human? Are all humans necessarily persons?
• Is a severely degraded human still a person? (For example, patients in a chronic vegetative state). When does a permanently degraded human lose personhood?
• What are the characteristics of an organism that are necessary or that are sufficient for an organism to be considered a person? Can animals be persons?
• Can intelligent, sentient machines be persons?
• What determines the "moral standing" of an organism?
• Does the degree of post-conception biological development of a human zygote determine whether or not that organism is a person?
• Does the genetic makeup of an organism determine whether or not the organism is human? (What about adding human genes to plants and animals?)
• What are the characteristics of an individual's genome that are necessary or that are sufficient for an organism to be considered a human?
• Do human enhancement technologies impact on whether or not an "enhanced" individual is still a person or a human?
• Is it possible to enhance an animal to the extent that they merit a change in moral standing?
• What of humans who have been "frozen" before death (cryonic suspension)?

issues associated with matters of humanness and personhood. Since each of these questions could ultimately be expanded into a complete treatise, the present effort will necessarily be limited. Driven both by personal curiosity as well as by the practical need for moral clarity as advanced medical technologies see clinical application, in this book I have chosen to focus on issues of personhood, moral standing, eugenics and human enhancement. I also consider the moral status of cryonically suspended individuals (most of which are "living" in the USA). Of these various interrelated topics, I have put special emphasis on the issues involved in pharmacologic enhancement, reflecting my view (a view based on observing many rapid developments in contemporary neuropharmacology) that the matter of pharmacologic enhancement represents one of the more pressing contemporary ethical challenges that awaits us.

Closely related to the discourse on what constitutes humanness or personhood are issues of moral standing, individual rights and ethical responsibility. For instance, it is commonly held that all individuals meeting personhood criteria are deserving of at least some individual rights and that only persons (and not, for instance, lower animals) are expected to be ethically responsible for their actions. Another ongoing philosophical debate concerns whether people (who may or may not be persons) deserve greater rights and higher moral standing than non-people; the rights of non-human animals are an example (Howard 2005; Sagoff 2007; Johnson 2013).

These issues are more than theoretical. For instance, the previously mentioned Great Ape Project (http://www.greatapeproject.org) is an international organization of animal advocates who are promoting the position that the United Nations should pass a "Declaration of the Rights of Great Apes" to confer basic legal rights to chimpanzees, bonobos, gorillas, and orangutans. These proposed rights would include the right to life, protection of individual liberty, release from captivity, and protection from torture.

1.7 Death

The philosophical issues associated with death are the subject of Chap. 5. This is a particularly important chapter that breaks new philosophical ground since it contributes substantially to the philosophical debate on where the line between life and death should be drawn.

Traditionally, death has been defined as the permanent cessation of the heartbeat and respiration. Modern developments in clinical resuscitation, however, have forced a reappraisal of this concept (Calixto and Leisman 2009). Today, in hospitals around the world, ventilators, dialysis equipment and drug infusions that sustain the circulation often permit the bodies of critically ill patients to be artificially supported despite severe physiological insults, including death of the brain itself. In addition, the advent of transplantation surgery has provided a strong clinical motivation to allow death to be defined in terms of loss of brain function (brain death) or in terms of a relatively brief period of asystole (cessation of cardiac activity) in some patients (non-heart-beating donors).

Issues such as these have led to animated discussion in the academic medical literature as the concept of death and brain death in particular can be said to be in evolution (Truog and Fackler 1992; Machado 1994; Egonsson 2009; Iltis and Cherry 2010). Finally, some thinkers argue that ongoing and future developments in cryonics and nanotechnology may require further reappraisal of when a person is really dead. Recent studies into the biochemistry of physiological (apoptosis) and pathological (necrotic) cell death in mammalian organisms have led to new insights into the concept of death on the micro scale (at the cellular level) as well as at the macro scale, i.e., the death of the entire biological organism.

As a result of these insights, some individuals have advanced the philosophical position that organismal death is not an event so much as it is a process, and that life and death are not distinct binary states, but entities with degrees and gradations. Thus, while a patient may be legally declared to be dead a few minutes after the continued cessation of cardiac activity (as is done, for example, with non-heart-beating organ donation protocols), nevertheless, these individuals argue, the cellular changes which occur in the period following cardiac arrest may be potentially reversible following the application of appropriate clinical interventions, especially under hypothermic conditions. Indeed, if all cells and organs in an organism died fully and completely immediately following the last heartbeat, any retrieved organs would be completely unsuitable for transplantation. These thinkers are now speculating that future developments in nanotechnology may permit repair and rejuvenation of organ tissue that is either dead or nearly dead by classical criteria (Merkle 1992; Hughes 2004).

With such issues in mind, some scholars have argued that it is helpful to distinguish between the definition of death based on cessation of the heart beat and a "more substantial" form of death, such as exists following cremation. In the latter case, sometimes called "absolutely irreversible death" or "information-theoretic death" destruction of the brain has occurred to such an extreme that any information it may have ever held is irrevocably lost for all eternity. This, some people argue, is the only real (irreversible) form of death.

1.8 Cryonics

Cryonics and suspended animation are the primary concern of Chap. 6. Like Chap. 5, it contributes substantially to the philosophical debate on where the line between life and death should be drawn.

The ultimate aim of cryonics is to achieve nondestructive freezing (cryopreservation) of advanced organisms like humans so that they can be safely thawed in the future, usually with a view to obtaining advanced medical treatment not currently available (Best 2009; Romain 2010). Although cryonics is usually dismissed by traditional cryobiologists as pseudoscience (Shermer 2001), citing intracellular ice crystal formation that permanently damages cellular structures, there is now evidence that such damage might eventually be mitigated through the use of cryoprotectants such as glycerol (Pegg 2007). In fact, advocates of cryonics take pains to point out that they do not actually want to freeze their patients, but rather subject them to "vitrification", a process that produces solidification of cellular materials in an amorphous glassy state that avoids cell-damaging ice formation (Vajta et al. 2009). While vitrification just prior to death from an illness would be expected to offer a far better chance of successful recovery ("thawing"), the clinically less desirable but legally more acceptable option of being vitrified a brief period after legal death is the method used in actual practice. As an example of ongoing advances in this field, in recent years Gregory M. Fahy has achieved successful vitrification and autotransplantation of a rabbit kidney through the use of a special vitrification solution (Fahy et al. 2009; Fahy and Wowk 2015).

1.9 Critique of Transhumanism

Chapter 7 is one that is particularly rich in philosophical arguments. The main focus is on dissecting the arguments against transhumanism that have been raised by a number of prominent philosophers and other individuals. Based on the critiques of transhumanism offered by critics such as Leon Kass, Francis Fukuyama, and Michael Sandel, I critically assess five commonly raised arguments why human enhancement by artificial means should not be undertaken: fairness, human dignity issues, "slippery slope" arguments, the "repugnance" argument, and safety issues. The weakness of "slippery slope" arguments and "repugnance" arguments are given special attention. In the end I conclude that safety issues are the most realistic concerns that the pursuit of transhumanism raises.

1.10 Transhumanism and Religion

Finally, Chap. 8 is concerned with the intersection of transhumanism, neurology and religious belief. The role of various brain structures in religious experiences (especially the frontal lobes) is discussed, including a discussion of how pathological brain conditions such as epilepsy can influence religious beliefs.

Chapter 2
Biomedical Ethics

Science cannot stop while ethics catches up—and nobody should expect scientists to do all the thinking for the country.

(http://thinkexist.com/quotation/science_cannot_stop_while_ethics_catches_up-and/160138.html) Elvin Stackman

Abstract This chapter is an outline of the basic issues in biomedical ethics and is intended to serve as a foundation for later chapters that draw on the concepts discussed. The chapter begins with an examination of the various categories of bioethical concerns as well as a look at the Hippocratic origins of medical ethics. Various instances of historical lapses in bioethics are then presented to the reader to serve as cautionary examples. The chapter then discusses various ethical theories, including Kantian ethics, principlism, virtue ethics and utilitarianism. Additional topics discussed include consent, medical futility, therapeutic privilege and a sampling of pertinent legal cases.

2.1 Introduction

Given that this work has as a central theme an exploration of the ethical issues related to life, death and transhumanism, an appropriate starting point would be to provide the reader with a brief background concerning bioethical concepts. Consequently, in this chapter a variety of topics related to the principles of bioethics is presented with the ultimate aim of establishing a contextual framework for the discussions presented in subsequent chapters. In particular, any discussion of the ethical propriety of transhumanist interventions (e.g. neurological enhancement via electronic devices, life extension via genetic engineering or cryonic preservation) is best preceded by a background discussing the various bioethical approaches available to systematically deal with these issues. In addition, a selection of material less directly germane to the central narrative of the book has been provided because of its central importance

D. J. Doyle, *What Does it Mean to be Human? Life, Death, Personhood and the Transhumanist Movement*, Anticipation Science 3,
https://doi.org/10.1007/978-3-319-94950-5_2

to bioethics. With this background the reader will be aided in his or her efforts to understand how the issues associated with the various debates related to life, death and transhumanism will necessarily depend on the conceptual "lens" through which these issues are explored.

Note that the discussion that follows, however, presents only the dominant positions in current bioethical discourse, and that a number of less well-established positions (e.g., feminist ethics, Marxist ethics, etc.) are not covered here. For information on these lesser known positions the reader is referred to standard sources (e.g., Singer 1996; Beauchamp and Childress 2013).

2.2 Overview of Ethics and Bioethics

Ethics is that branch of philosophy which seeks to address questions about proper conduct, rightness and wrongness, goodness and evil, virtue, honor, integrity, justice and related notions. It is often divided into a number of overlapping branches. **Applied ethics** is concerned with the problem of how ethical outcomes might be achieved in real-world situations. Applied ethics includes subfields such as business ethics, military ethics, judicial ethics, bioethics, and professional ethics. **Descriptive ethics** is concerned with documenting and interpreting what moral principles people actually abide by. The famous work of Kohlberg (1971) is an example of this field, although his work also overlaps with the field of **moral psychology**. Descriptive ethics stands in contrast with **normative ethics**, which is concerned with ethical theories that *prescribe* how people *ought* to act (as opposed to merely describing how people actually act.) Normative ethical theories include deontological approaches (e.g., Kant's Categorical Imperative), consequentialist approaches, (e.g., utilitarian ethics), virtue ethics, and other theories. Finally, according to the Stanford Encyclopedia of Philosophy, **meta-ethics** "is the attempt to understand the metaphysical, epistemological, semantic, and psychological, presuppositions and commitments of moral thought, talk, and practice" including such questions as "Is morality more a matter of taste than truth? Are moral standards culturally relative? Are there moral facts? If there are moral facts, what is their origin?"[1]

A number of arbitrary overlapping subdivisions of bioethics can be envisioned.[2] **Clinical bioethics** is concerned with the day-to-day ethical decisions clinicians such as physicians and nurses encounter in caring for patients. Clinical bioethics may itself be subdivided into several subfields, such as nursing bioethics, pain management bioethics, critical care bioethics, surgical bioethics, genetics bioethics, neonatal bioethics, anesthesiology bioethics, biomedical engineering ethics, etc. Some examples may be helpful to clarify these subdivisions.

---

[1] http://plato.stanford.edu/entries/metaethics/.

[2] Although this particular categorization is my own, it has undoubtedly been influenced by both my prior classroom exposure to bioethics as well as by my reading of various canonical texts (e.g., Beauchamp and Childress 2013).

An example of a nursing bioethical issue concerns the appropriateness of nurses participating in judicial executions in a technical role such as establishing vascular access for lethal injection. An example of a pain management bioethical issue concerns the appropriateness of being "generous" in the administration of potentially addictive opiates such as morphine in patients with severe nonmalignant pain syndromes. An example of an anesthesiology bioethical issue concerns the appropriateness of aggressively resuscitating terminally ill DNR ("Do Not Resuscitate") patients who are doing poorly while undergoing palliative care surgery. An example of a biomedical engineering ethical issue concerns the safety testing requirements for a new medical product awaiting clinical deployment, for example, a new drug infusion pump.

A second major division of bioethics, **foundational bioethics**, is concerned with the philosophical foundations of bioethics. This field is concerned with issues such as the merits of the deontological approach versus the utilitarian approach to approaching ethical issues, or the problem of handling an ethical issue when a conflict in ethical principles arises. Note, however, that not all bioethical approaches are principle-based; some approaches emphasize the moral context over categorical principles (e.g., situational ethics, feminist ethics).

A third division of bioethics, **cultural bioethics**, is concerned with ethical questions in relation to cultural and social settings. Ideological, religious, and historical dimensions often influence the discourse here. For instance, some feminists advocate an approach to bioethics emphasizing an ethic of caring where human relationships, feelings and emotions are more pertinent to making ethical decisions than traditional top-down rationalistic approaches based on abstract philosophical principles (most of which have been developed by male philosophers) (Sherwin 1992).

Some examples of cultural bioethical questions may help here. Is male circumcision a form of genital mutilation? Should patients who are being clinically harmed by their cultural or religious beliefs be provided with information that is critical of their beliefs with a view to changing their "toxic" belief system? Should a husband have any say over the medical care his wife is to receive? What should be the role of government in the delivery of health care? What philosophical principles might a Marxist-Leninist physician practicing in North Korea use as the basis for providing ethical care to her patients?

Finally, **regulatory and policy bioethics** is concerned with ethical issues such as clinical trial design, drug approval ethics, use of embryonic tissue in research and clinical treatment, human cloning, organ procurement for transplantation, xenotransplantation, assisted suicide, euthanasia, formulation of brain death policy, allocation of scarce health care resources, etc.

I SWEAR by Apollo the physician, and Aesculapius, and Health, and All-heal, and all the gods and goddesses, that, according to my ability and judgment, I will keep this Oath and this stipulation - to reckon him who taught me this Art equally dear to me as my parents, to share my substance with him, and relieve his necessities if required; to look upon his offspring in the same footing as my own brothers, and to teach them this art, if they shall wish to learn it, without fee or stipulation; and that by precept, lecture, and every other mode of instruction, I will impart a knowledge of the Art to my own sons, and those of my teachers, and to disciples bound by a stipulation and oath according to the law of medicine, but to none others. I will follow that system of regimen which, according to my ability and judgment, I consider for the benefit of my patients, and abstain from whatever is deleterious and mischievous. I will give no deadly medicine to any one if asked, nor suggest any such counsel; and in like manner I will not give to a woman a pessary to produce abortion. With purity and with holiness I will pass my life and practice my Art. I will not cut persons laboring under the stone, but will leave this to be done by men who are practitioners of this work. Into whatever houses I enter, I will go into them for the benefit of the sick, and will abstain from every voluntary act of mischief and corruption; and, further from the seduction of females or males, of freemen and slaves. Whatever, in connection with my professional practice or not, in connection with it, I see or hear, in the life of men, which ought not to be spoken of abroad, I will not divulge, as reckoning that all such should be kept secret. While I continue to keep this Oath unviolated, may it be granted to me to enjoy life and the practice of the art, respected by all men, in all times! But should I trespass and violate this Oath, may the reverse be my lot!

Fig. 2.1 The Original Hippocratic Oath, translated from ancient Greek. *Source* MIT tech classics collection at http://classics.mit.edu/Hippocrates/hippooath.html

2.3 Hippocratic Origins of Bioethics

In antiquity, Hippocrates (c. 460 BCE–c. 370 BCE), a profoundly influential Greek physician born on the Greek island of Cos, developed a professional oath for his student physicians (Davey 2001) (Fig. 2.1). This was one of the first professional codes of ethics, and remains important to this day.

Although not all of Hippocrates' teachings have stood the test of time (for instance, he taught that illness resulted from an imbalance of the four humors in the body: blood, black bile, yellow bile and phlegm), his ethical precepts have left an important legacy, and a modified version of his oath is often taken by medical students prior to graduation from medical school.

As is evidenced in Fig. 2.1, the original Hippocratic Oath forbids a number of activities, such as abortions, bladder stone surgery, euthanasia, disclosure of confidential information without the patient's permission, and sex with patients (and their slaves!). Obviously, some of these rules have changed over time and are no longer appropriate to contemporary medical practice (for example, the rule against bladder stone surgery would upset many urologists), which explains why medical students today usually take a modified Hippocratic Oath (or an entirely different oath) rather than the original Hippocratic Oath (Friedlander 1982; Kao and Parsi 2004).

Since the time of the Hippocratic Oath, many variations have been promulgated in an attempt to deal with some of the difficulties identified above, as well as to adapt to changing medical practices and cultural circumstances. We briefly discuss some of these variations below both to show how medical ethics thinking has evolved over time as well as to provide context for discussions to follow.[3]

The World Medical Association (WMA) has published their version of the Hippocratic Oath in October 1949. Known variously as the *Declaration of Geneva* or the *WMA International Code of Medical Ethics*, the most recent revision, published in 2006, can be viewed online at http://www.wma.net/en/30publications/10policies/c8/index.html. It is notably different from the original Hippocratic Oath in that no mention is made of prohibitions to conducting abortions, bladder stone surgery, or (surprisingly) euthanasia.[4]

The World Medical Association has also issued a companion policy document widely known as the *Declaration of Helsinki*. (World Medical Association 2002; Diamant 2002; Puri et al. 2009). It is a set of ethical principles for the medical community regarding human experimentation and is widely regarded as the ethical foundation for human research.[5]

As a consequence of these and other developments, in 1979 the USA has issued its own policy document on the ethical conduct of research involving human subjects. Known as the *Belmont Report* after the conference center where the document was drafted, it is the basis for the complex regulatory framework that governs human subject research in the USA (Cassell 2000; Smith 2001; Marshall 1986; Beauchamp 2004).[6]

Although the Belmont Report is central to American bioethics policy, it has not escaped criticism. Criticisms include concerns that the approach is "one size fits all" in nature and that the underlying ethical analysis fails to take into account potentially important factors, such as racial, ethnic, cultural, gender, and geographical considerations (Miller 2002).

---

[3]The reader may note, for example, a shift from a set of specific governing rules (e.g., from the Hippocratic oath: "I will not cut persons laboring under the stone") to a set of broad-based governing principles (e.g., from the Nuremberg Code: "The voluntary consent of the human subject is absolutely essential") as one moves from the original Hippocratic Oath and its variants to more recent ethical guidelines such as the Nuremberg code, the Declaration of Helsinki, and the Belmont Report.

[4]I use the word "surprisingly" because the prohibition against euthanasia is a long-standing, central tenet of classical medical ethics.

[5]The most recent revision, issued in 2008, may be viewed online at http://www.wma.net/en/30publications/10policies/b3/index.html.

[6]The report may be viewed online at http://ohsr.od.nih.gov/guidelines/belmont.html.

2.4 Lapses in Medical Ethics I: The Origins of Formal Bioethics

During World War II a sizable number of Nazi physician-scientists performed medical experiments using concentration camp prisoners at Auschwitz, Buchenwald, Dachau, and Sachsenhausen (Barondess 1996). While a great many of these experiments were worthless amateur efforts, reflecting a lack of planning and discipline, some of the experiments were comparatively well-executed (from a strictly amoral perspective) and resulted in potentially useful scientific information. Regardless, all these experiments were highly unethical. First, nearly all the experimental subjects endured tremendous unnecessary suffering and distress. Second, the experimental protocols were usually designed with a fatal outcome as the end-point (primarily to allow for post-mortem examination). And third, the participants were all enrolled against their will (Lifton 1986).

Following the war, twenty Nazi doctors were charged with War Crimes and Crimes Against Humanity before the International Military Tribunal at Nuremberg, and it was there that the full horror of their shameful endeavors came to the attention of the world (Katz 1996). At the tribunal one of the more pragmatic experimental programs to be identified concerned developing possible means to mitigate the clinical response to hypothermia, an issue dear to the heart of Luftwaffe pilots, who were known to occasionally crash land into the cold North Sea following enemy fire. In response to this need, and with the explicit support of Reichsführer Heinrich Himmler, Dr. Sigmund Rascher, a physician as well as a Luftwaffe officer, initiated a research program at Dachau in which prisoners were immersed in ice water baths. While immersed, rectal temperature, heart rate, the level of consciousness and the degree of shivering were meticulously recorded. In all, 300 individuals were sacrificed (Berger 1990). Perhaps the most important thing learned was that victims of hypothermia will usually die (from ventricular fibrillation it turns out) when the rectal temperature drops below 25 °C (77 °F).

Not all the test subjects died, at least not immediately, since some of the rewarming methods studied were evidently quite effective. While immersion in warm water proved to be one highly effective method (and remains in use to this day), the experimenters—allegedly at the suggestion of Himmler himself—even placed some male test subjects in a warm bed with a naked woman (also a camp prisoner) to explore the re-warming potential of sexual intercourse.[7]

One ethical issue that arose decades later with the Nazi hypothermic data concerned the ethics of incorporating this information, previously unpublished, into a scientific review of the topic. This issue faced the editor of the *New England Journal of Medicine* in the late 1980s when then editor Arnold Relman, ultimately refused permission to publish parts of the Nazi hypothermia data as a component of a potentially life-saving research report from the Hypothermia Laboratory at the University

[7] http://www.bordeninstitute.army.mil/published_volumes/ethicsVol2/Ethics-ch-15.pdf.

of Minnesota (Angell 1990).[8] Naturally, such a position has the unfortunate consequence that potentially important information that may help patients may end up being denied to these individuals and, furthermore, the process of repeating the experiments so that they are in full compliance with regulations may expose new experimental subjects to new risks so as to obtain information that was already available. As a result, some ethicists favor a policy for such cases where the paper in question is accepted, but is published with an accompanying editorial in which the ethical aspects of the manuscript are critically discussed.

One very important consequence of the horrible Nazi medical experiments was the recognition that there existed a pressing need for a formal code of ethics to guide medical research involving human subjects. In fact, during the Nuremberg trials some of the accused had argued that there was no law or code that differentiated between legal/ethical and illegal/unethical medical experiments.[9,10] The result was a ten-point policy statement that constituted the "Nuremberg Code" (Kious 2001; Temme 2003; Becker 2005; Lemaire 2006; Adam 2007) (Fig. 2.2).

Since the time of its launching, the Nuremberg Code has had a profound influence on biomedical ethics, establishing such principles as informed consent for experimental procedures (and by extension, for ordinary clinical procedures), the need to ensure against force and coercion, the need for properly planned and executed scientific experimentation, and the need to protect the interests of experimental participants.

For the most part, the code has stood up under changing times and circumstances to serve as a timeless guide to biomedical research ethics (Shuster 1997).[11]

[8]Relman was probably less concerned about the quality of the experimental data (although the project's chief investigator, Sigmund Rascher, was admittedly a clinician and not a qualified scientist) than about the ethical propriety of publishing experimental data obtained in clear violation of the Nuremberg Code and the traditions of the journal, which had earlier published a sizeable number of seminal articles in the field of bioethics. For an interesting discussion on this matter, the reader is directed to a commentary by Berger (1990).

[9]Despite the absence of any code of ethics, the Nazi experiments were considered unethical on the principal grounds that the experiments were not conducted with the consent of the participants. Some experiments were judged to be unethical because there was an a priori reason to believe that either death or a significant injury would occur. Some experiments (such as those conducted by Dr. Joseph Mengele) were additionally unethical because they stood no chance of providing clinically useful information despite posing substantial risks and suffering to the experimental subjects.

[10]It is interesting to ask whether there would have been a need for a formal code of ethics to guide medical research if the Holocaust experiments had never taken place. My answer is that ethical codes such as the Nuremberg Code would still have arisen (albeit perhaps not as historically early) both because other identified ethical lapses in medical research would have made the need apparent, as well because of a perceived need by ethically sensitive medical researchers dealing with increasing complex experimental paradigms.

[11]Some authorities argue that the Nuremberg Code's requirement that animals be used in lieu of humans in the early phases of research and testing are based on scientifically outdated principles, and serves no useful function. One basis for this position concerns important genetic differences in how drugs are handled across various species. A specific example is the drug isoproterenol, once used to treat asthma, which turned out to be harmful to humans in the amounts originally recommended based on animal studies. Authors such as Greek et al. (2012) argue that the Nuremberg Code's

NUREMBERG CODE

1. The voluntary consent of the human subject is absolutely essential. This means that the person involved should have legal capacity to give consent; should be so situated as to be able to exercise free power of choice, without the intervention of any element of force, fraud, deceit, duress, over-reaching, or other ulterior form of constraint or coercion; and should have sufficient knowledge and comprehension of the elements of the subject matter involved as to enable him to make an understanding and enlightened decision. This latter element requires that before the acceptance of an affirmative decision by the experimental subject there should be made known to him the nature, duration, and purpose of the experiment; the method and means by which it is to be conducted; all inconveniences and hazards reasonable to be expected; and the effects upon his health or person which may possibly come from his participation in the experiment.

The duty and responsibility for ascertaining the quality of the consent rests upon each individual who initiates, directs or engages in the experiment. It is a personal duty and responsibility which may not be delegated to another with impunity.

2. The experiment should be such as to yield fruitful results for the good of society, unprocurable by other methods or means of study, and not random and unnecessary in nature.

3. The experiment should be so designed and based on the results of animal experimentation and a knowledge of the natural history of the disease or other problem under study that the anticipated results will justify the performance of the experiment.

4. The experiment should be so conducted as to avoid all unnecessary physical and mental suffering and injury.

Fig. 2.2 The Nuremberg Code. Reprinted from Trials of War Criminals before the Nuremberg Military Tribunals 208 under Control Council Law No. 10, Vol. 2, pp. 181–182. Washington, D.C.: U.S. 209 Government Printing Office, 1949. See also http://ohsr.od.nih.gov/guidelines/nurem210 berg.html.

---

requirement for animals testing only serves to "increase the cost of drug development, and prevent

5. No experiment should be conducted where there is an a priori reason to believe that death or disabling injury will occur; except, perhaps, in those experiments where the experimental physicians also serve as subjects.

6. The degree of risk to be taken should never exceed that determined by the humanitarian importance of the problem to be solved by the experiment.

7. Proper preparations should be made and adequate facilities provided to protect the experimental subject against even remote possibilities of injury, disability, or death.

8. The experiment should be conducted only by scientifically qualified persons. The highest degree of skill and care should be required through all stages of the experiment of those who conduct or engage in the experiment.

9. During the course of the experiment the human subject should be at liberty to bring the experiment to an end if he has reached the physical or mental state where continuation of the experiment seems to him to be impossible.

10. During the course of the experiment the scientist in charge must be prepared to terminate the experiment at any stage, if he has probable cause to believe, in the exercise of the good faith, superior skill and careful judgment required of him that a continuation of the experiment is likely to result in injury, disability, or death to the experimental subject.

Fig. 2.2 (continued)

2.5 Lapses in Medical Ethics II: The Tuskegee Syphilis Study

Although the Nazi medical experiments are perhaps the best-known collection of medical ethics violations, it is important to emphasize that there have been a great many others. For instance, in the USA a serious lapse in medical ethics took place that had an impact arguably comparable to that of the Nazi medical experiments. This was a clinical study conducted by the U.S. Public Health Service between 1932 and

---

otherwise safe and efficacious drugs and therapies from being implemented." In addition, some authorities argue that, in this particular respect, the Nuremberg Code fails to consider the moral rights of animals. For additional insight into this interesting and important problem see Shanks et al. (2009). For a comprehensive discussion on the moral rights of animals see Singer (1975).

1972 in Tuskegee, Alabama. Investigators recruited African-American sharecroppers to study the natural progression of untreated syphilis. While no effective treatment was available when the study was launched, a serious ethical lapse took place when in the late 1940s penicillin become available as an effective treatment for syphilis, yet the Tuskegee scientists continued the study, withholding penicillin treatment and information about it from the patients (Katz et al. 2008a, b; White 2006a, b).

Victims of this moral misadventure included study participants who died of syphilis, sexual contacts who contracted the disease, family members who cared for the infected individuals, and children born with congenital syphilis.

The study ended suddenly in 1972 when the leaked story was reported in the *Washington Star* on July 25, 1972 and became front-page news in the *New York Times* the following day. The US Centers for Disease Control, which was responsible for advocating the continuation of the study in 1969 despite numerous legal and ethical objections, offers further details online at http://www.cdc.gov/tuskegee/index.html.[12]

2.6 Lapses in Medical Ethics III: Japanese Wartime Medical Atrocities

During World War II Japanese physicians and researchers performed barbarous biological experiments on living humans in programs such as Unit 731 (Harris 1992; Watts 2002). For instance, some prisoners were injected with pathogens to study their effects, while others were used to test flame throwers and grenades positioned at various distances and positions.

In a second breach of ethics, acting in "the national interest", at the end of the war the US government secretly granted immunity from war crimes prosecution to the perpetrators (Dickman 1988; Chen 1997; Nie 2006; Blumenthal and Miller 1999).

2.7 Ethical Theory

One of the first philosophers to systematically explore the philosophical foundations of ethics was Aristotle (c. 384 BCE–c. 322 BCE), famous for his *Nicomachean Ethics* and numerous other works (Allmark 2006). Philosophers who have additionally con-

---

[12]While justification for continuing the study following the availability of effective treatment is difficult to envision, the arguments offered were that continuing the study would provide information about the natural history of syphilis not readily available by other means, that the participants were receiving free medical care that they would likely not otherwise receive, and that the study might even show the pharmacologic treatment of syphilis to be unnecessary. Some physicians, believing that only autopsies would provide definitive information about the natural history of syphilis, lacked an interest in seeing their experimental subjects survive.

tributed greatly to ethical theory include Thomas Aquinas, Immanuel Kant, Baruch Spinoza, Jeremy Bentham, John Stuart Mill and a great many others.

Ethical theories are the foundations upon which ethical analysis is conducted; they are the viewpoints from which moral guidance can be obtained. Each ethical theory emphasizes a different aspect of ethical decision making in reaching an ethically "correct" decision (Beauchamp and Childress 2013; Lawlor 2007).[13]

One approach to morality is to focus on the consequences of a particular moral action. The ***consequentialist*** approach determines moral responsibility by weighing the consequences of one's actions (Beauchamp and Childress 2013). According to the consequentialist view, correct moral actions are determined by careful deliberation concerning the anticipated consequences (e.g. costs, benefits, preferences etc.) of an action for the various stakeholders involved. Several subtypes of consequentialism have been proposed: (1) the view that an action is morally correct if its consequences are more positive or favorable than negative to the person performing the action (*ethical egoism*); (2) the view that an action is morally correct if the consequences of that action are more positive than negative to everyone except the person doing the action (*ethical altruism*); and (3) the view that an action is morally correct if the action's consequences are more positive than negative to everyone (*utilitarianism*). Utilitarianism can be further subdivided, as is discussed in the next paragraphs.

Act utilitarianism takes the position that, when facing a moral choice, one must consider the expected consequences of various potential actions and, based on this analysis, choose to do what we believe will generate most happiness or pleasure for the most people. A rule utilitarian, by contrast, analyses a moral dilemma by looking at potential rules of action that may be applicable, and adheres to the rule that would be expected to produce the most happiness or pleasure. Negative utilitarianism requires us to act so as to produce the least amount of evil or harm for the greatest number of people. In the case of preference utilitarianism, advocated by Professor Peter Singer (Jamieson 1999), the goal is to meet the preferences of the greatest number of people. This is discussed in Sect. 2.11.

Another approach to morality is to focus on duties and responsibilities instead of consequences. The ***deontological*** *approach* to morality (from the Greek word *deon*, meaning *obligation* or *duty*) is based on specific obligations or duties. These can be either positive/prescriptive (such as to care for our family) or negative/proscriptive (such as not to steal). This approach is also sometimes called *nonconsequentialist*

[13] In the present treatment the terms "moral" and "ethical" are used interchangeably, as per the practice of a number of authors. A number of other authors, however, clearly distinguish between the two terms, where ethics is taken to be that branch of philosophy which addresses questions about morality, while morality is taken to signify the moral principles of a particular tradition, group, or individual (http://plato.stanford.edu/entries/morality-definition/). In addition, some authors would restrict the use of the term "moral" to philosophical systems such as that developed by Kant, based on notions such as duty, obligation, and principles of conduct. Other authors state that ethics refer to the series of rules provided to an individual by an external source, such as their profession, while morals refer to an individual's own principles regarding right and wrong. A distinguishing example sometimes offered is that a lawyer's personal morals may tell him that murder is reprehensible and that murderers should be punished, but his professional ethics as a defense attorney may require that he defend his client with enthusiasm, even if he suspects that the client may actually be guilty.

since these principles are held to be obligatory regardless of any good or bad consequences that might result. For example, it is wrong to kill even if it results in great benefit.

Other deontological approaches to ethics are "duty theory" popularized by Ross and "rights theory" (concerned with rights that all people have, and which the rest of us must respect) (Beauchamp and Childress 2013). Ross's duty theory defines duties of beneficence, non-malfeasance, justice, self-improvement, reparation, gratitude and promise-keeping. He calls these *prima facie* duties. This approach was developed as an alternative to utilitarianism because of perceived failures of utilitarianism as a satisfactory moral theory:

> [Utilitarianism] seems to simplify unduly our relations to our fellows. It says, in effect, that the only morally significant relation in which my neighbours stand to me is that of being possible beneficiaries of my action. They do stand in this relation to me, and this relation is morally significant. But they may also stand to me in the relation of promisee to promisor, of creditor to debtor, of wife to husband, of fellow countryman to fellow countryman, and the like; and each of these relations is the foundation of a... duty. (Ross 1930)

As noted above, rights theory is concerned with rights that all humans have, and which other humans must respect. A right is a benefit (e.g., a liberty, a power, a prerogative, an immunity) that someone gains by virtue of his or her particular status as a citizen, a human being, a woman, a man, a child, a minority, a sentient animal etc. Rights can be positive, such as rights to food, clothing, and shelter, or negative, such as the right to be left alone. Rights theory thus offers an approach to moral action in that actions that violate the rights of others are said to be immoral.

As can be seen from the above discussion, a number of approaches to tackling ethical issues are available. In the following sections we elaborate on some of the points made above.

2.8 Kantian Ethics

Immanuel Kant (22 April 1724–12 February 1804) was a profoundly influential German philosopher of the Enlightenment. His work *The Critique of Pure Reason* (1781) is often cited as one of the most significant works of modern philosophical history while his *Groundwork for the Metaphysics of Morals* (1785) explains his views on ethics and morality. Fundamental to Kantian ethics is the principle ("categorical imperative")[14] that we must always treat people as ends in themselves, and never only as a mere means to an end. By this Kant intended that we should always treat

[14]A *categorical imperative* is an absolute, unconditional moral requirement that applies to all circumstances. It is distinguished from a *hypothetical imperative*, where certain ends apply (e.g., if I wish to tame my hunger, I must eat a meal). Note that while it is common to identify the categorical imperative as a principle, it is easy to get the wrong idea about what Kant takes it to be. Kant's *categorical imperative* is a command, an unconditional command, and as such is neither true nor false. And although its form can be taken to be a self-generated, self-directed command, it is not an object of knowledge. It is essentially *practical*—something we *do*.

people with humanity and dignity, and never use individuals as "mere instruments" as a means to further our own happiness (Gensler 1985; Morelli 1999; Gunderson 2004; Bernstein and Brown 2004; Heubel and Biller-Andorno 2005). Another way of expressing Kant's Categorical Imperative is: "Always act in such a way that the maxim of your action can be willed as a universal law." In other words, one should act only in such a way that one would want all persons to act, and not treat individuals as mere "means to an end." Kant's moral philosophy is an example of a deontological approach[15] to ethics (vide infra).

Related to Kantian ethics is the concept of a maxim. Simply stated, a maxim is any simple and memorable rule or guide for living, as with Shakespeare's maxim from Hamlet: 'neither a borrower nor a lender be'.[16] Kant saw maxims as a "principle of action" in the sense that in Kantian ethics, the categorical imperative allows one to test candidate maxims to determining whether the actions they refer to are morally permissible.

In his *Critique of Practical Reason*, Kant provides an example of a maxim and of how to apply the categorical imperative test: "I have, for example, made it my maxim to increase my wealth by any safe means. Now I have a deposit in my hands, the owner of which has died and left no record of it.... I therefore apply the maxim to the present case and ask whether it could indeed take the form of a law, and consequently whether I could through my maxim at the same time give such a law as this: that everyone may deny a deposit which no one can prove has been made. I at once become aware that such a principle, as a law, would annihilate itself since it would bring it about that there would be no deposits at all."[17]

2.9 Principlism

One commonly used approach to tackling bioethical problems in the Western world is to invoke the guiding principles of the "Georgetown School" of bioethics, a philosophical system with origins in the University of Georgetown that has become popular and profoundly influential by virtue of its suitability for addressing many real-world bioethical problems.

The Georgetown School calls upon four ethical principles (Beauchamp and Childress 2013). These are [1] *respect for autonomy* (the right to actively participate in medical decisions concerning oneself without being dictated to or controlled by other parties); [2] *beneficence* (the requirement that caregivers, all else being equal, should do what they can to improve the patient's situation); [3] *non-malfeasance* (the requirement to avoid bringing harm to the patient); and [4] *justice* (requiring

---

[15]As discussed herein, deontological approaches to ethics judge the ethical propriety of an action based on the action's adherence to one or more rules.

[16]Hamlet, Act 1, scene 3, 75–77.

[17]Kant, Immanuel; translated by Mary Gregor [1788] (1997). *Critique of Practical Reason*. Cambridge University Press. pp. 25/5:27. ISBN 0-521-59051-5.

the fair and impartial treatment of all persons, especially in the context of resource allocation).

These principles, as discussed in the Belmont Report on the protection of human research subjects (discussed above), are also the foundation for U.S. federal regulations that govern clinical research [e.g., Code of Federal Regulations; Protection of Human Subjects 45 CFR 50 and Code of Federal Regulations; Food and Drugs; Institutional Review Boards 21 CFR 56.]

Of these four bioethical principles, the principle of autonomy is one that is central to a good deal of contemporary bioethical discourse and appears as an important issue in a number of present-day bioethical debates, such as therapeutic abortion, voluntary euthanasia, medical research on human subjects, and the right of patients to refuse clinically necessary medical treatments.

It should be emphasized that Principlism has its critics. For example, in a paper entitled "In praise of unprincipled ethics", Harris (2003) makes the case against Principlism as a basis for biomedical ethics, stating that "while the four principles constitute a useful "checklist" approach to bioethics for those new to the field….it is an approach which if followed by the bioethics community as a whole would… lead to sterility and uniformity of approach of a quite mindbogglingly boring kind."[18] He goes on to state that "while in ethics generally and bioethics in particular, doing good, avoiding harm, protecting justice, and respecting rights and interests are never likely to be far from our minds, they are neither necessarily the best nor the most stimulating way of approaching all bioethical dilemmas."

Other concerns have been raised by Clouser and Gert (1990), where they argue that the underlying principles in Principlism "lack any systematic relationship to each other, and they often conflict with each other" and that these conflicts "are unresolvable, since there is no unified moral theory from which they are all derived." They conclude that in using principles to replace both general moral theory and particular moral rules in handling the moral conundrums that arise in clinical practice, Principlism fails to provide a useful framework for medical ethics.

In a later work Clouser (1995) elaborates on his concerns with Principlism. Here it is argued that "unlike the principles of Kant, Mill, and Rawls" the guiding principles behind Principlism "are not action guides that stem from an underlying, integrated moral theory" with the result that problems inevitably arise in reconciling the principles with each other, such as conflicts that sometimes arise between the principle of "respect for autonomy" and the beneficence principle. Additionally, Clouser sees

[18]It is not clear why uniformity of approach to tackling bioethical problems should be a bad thing. At a minimum a uniform approach would be expected to produce more consistent outcomes compared to an approach that draws on different philosophical bases on different days. The criticism that Principlism is "sterile" is arguably meaningless (at least in the sense used by the logical positivists), but in any event fails to appreciate the real value that it so often offers in dealing with the complex, real-world bioethical problems encountered in real life. While it is undoubtedly true that the central tenants in Principlism sometimes come into conflict (most notably, *respect for autonomy* versus *beneficence*) it is not clear that other approaches to dealing with ethical issues are free of similar difficulties. For example, both the deontological and consequentialist approaches to ethics have important limitations that are discussed later in this chapter.

difficulty in interpreting the elements of Principlism as guides to moral action "since they have no content in and of themselves." He goes on to state that (p. 223):

> … each principle [within Principlism] functions mostly as a reminder that there is an ethical value that the agent ought to take into account—the principle does not tell the agent what or how to think, or how to deal with the value in a particular instance—but it reminds him to consider it. So he does. But receiving no guidance, he deals with it in his own way, gives it his own weight, gives it his own interpretation, and grants his own exceptions—and then, as validation of his position and conclusion, he cites the relevant principle—which in effect only says something like "Think about justice" or "Think about helping people."

Clouser offers us the following additional commentary (p. 224) that reflects some of the concerns that are discussed later in this chapter (Sect. 2.22):

> The use of the principles of principlism seems to accentuate what Gert and I have called the "anthology syndrome" in bioethics. It is a kind of relativism espoused (perhaps unwittingly) by many books (usually anthologies) of bioethics. They parade before the reader a variety of "theories" of ethics—Kantianism, deontology, utilitarianism, other forms of consequentialism, and the like—and say, in effect, choose whichever of the competing theories, maxims, principles, or rules suits you for any particular case. Just take your choice! They each have flaws—which are always pointed out—but on balance, the authors seem to be saying, they are probably all equally good!
>
> Similarly the principles of principlism are unconnected with each other, and, although each embodies the key concern from one or another theory of morality, there is no account of how they should relate to each other. One just chooses the principle that appears most important in any particular case. Thus principlism obscures and confuses moral reasoning by its failure to provide genuine action guides and by its not having an underlying moral theory.

Beauchamp (1995) has offered a rebuttal to those who advocate different approaches[19] to moral problem solving, stating that while "often presented as rival methods or theories, these approaches are consistent with and should not be considered adversaries of a principle-based account."

2.10 Ethical Conflicts in Principlism: Autonomy as "First Principle Among Equals"

As noted above, it is sometimes possible that some of the principles in a system of morality come into conflict with each other. For example, in the case of Principlism, ethical difficulties may arise when a competent adult patient's desires conflict with a caregiver's beneficent duty to work for the best clinical interests of the patient. The best-known example of this is the Jehovah's Witness patient who refuses a clinically necessary blood transfusion or a patient with severe peripheral vascular disease who refuses to quit smoking. In Western culture, it is held that as long as the patient is a competent adult and understands the consequences of the decision, then the physician

[19] Examples of different approaches would include, for example, Impartial Rule Theory, Casuistry (Jonsen 1995), and (3) Virtue Ethics (Pellegrino 1995).

should respect the patient's decision despite the fact that the decision is unwise and imprudent.

However, not everyone is comfortable with such a position. For example, in countries like China, where personal autonomy is often taken to be less important than the needs of the state, one could imagine bioethicists offering a different position. For anesthesiologists dealing with Jehovah's Witness patients another dilemma can sometimes arise: because many Jehovah's Witness patients are not likely to be fully informed about possible inconsistencies in church doctrine regarding medical issues, should anesthesiologists attempt an in-depth discussion to encourage a rational and truly autonomous decision (for instance, using information provided by the Jehovah's Witness "reform" movement [www.ajwrb.org]), or should they remain silent regardless of the degree to which they might find the patients' views to be misguided, so as to avoid the perception of "religious interference"?

2.11 Peter Singer

We saw earlier (Sect. 2.7) that classical utilitarianism advocates the principle of providing "the greatest happiness to the greatest number" as the basis for assessing the morality of various actions. Over the years utilitarianism has undergone a number of refinements, such as "act utilitarianism," "rule utilitarianism," "negative utilitarianism" and "preference utilitarianism" (Gillon 1985).

Peter Singer, professor of bioethics at Princeton University's Center for Human Values, is a profoundly influential contemporary philosopher who espouses rationalist, atheist, and utilitarian viewpoints. Within the latter category he is known as an advocate of "preference utilitarianism." This form of utilitarianism is rather different from the classic definition of providing the greatest happiness to the greatest number of people. Because of the difficulty of measuring happiness, his approach aims at satisfying the "preference" of those affected. A moral decision thus amounts to determining the strongest preferences of all those affected by the decision. This approach has led him to articulately challenge closely held beliefs on infanticide, euthanasia, and the moral status of animals (Singer 1975, 1996, 2009; Jamieson 1999).

Two fundamental ethical precepts are at the heart of Singer's philosophy. Firstly, one should not inflict unnecessary pain on any living thing; and secondly, the defining characteristic of a "person" is self-awareness. Singer would thus see it as ethically correct to allow the killing of unwanted neonates born with severe neurological disabilities. He would argue that they are not self-aware and therefore not persons. Since neonates can feel pain, Singer's first ethical precept dictates that this killing be done as painlessly as possible—for example, by lethal injection. He would see this as more humane than allowing death through withdrawal of care (Jamieson 1999; Singer 1996; Howard 2005).

Singer advocates a "quality-of-life" ethic as a replacement for the more traditional "sanctity-of-life" framework advocated by many established moral philosophers and religious leaders. In particular, he is a proponent of a "journey model of life", which establishes the wrongness of killing by the degree to which this would frustrate a person's (not necessarily a human person's)[20] life goals. He also holds that a person's right to life is fundamentally tied to that person's mental capacity to have preferences, to feel pain and to experience pleasure.

As a result of this latter position, for example, Singer offers a utilitarian approach to abortion that weighs the preferences of a mother against the preferences of the fetus. Since a fetus, at least in the early weeks of pregnancy, has no capacity to think, to suffer or to hold preferences, there is nothing to weigh against a mother's preferences to undergo an abortion.[21] This makes Singer see abortion as morally permissible.

Such examples demonstrate a surprising openness and frankness of thought that makes Singer's opinions unusual in their clarity and their transparency—Singer studiously avoids euphemisms that muddy the waters, such as calling a "killing" something other than what it is. This makes his ethical philosophy clear and direct even if it is at times counterintuitive.

In the realm of animal rights, Singer argues that human beings do not have primacy over other species merely because they are human. He regards such notions as the last bastions of pre-Darwinian (or pre-Copernican) thought that once permitted slavery. Singer argues that increasing moral awareness and ethical sensitivity mean that future generations will see our current treatment of animals with the moral outrage we now reserve for slavery. This has obvious implications for researchers using animal models (Singer 1975).

2.12 A Transplantation Example

A clinical example adapted from Thomson (1985) may help explain utilitarianism. Consider a transplantation specialist with four patients. One needs a new liver, one needs a set of lungs, one needs a heart, and one needs a kidney. An act utilitarian would theoretically be comfortable with the idea of hunting down and kidnapping the first healthy person he encounters with a view to using him as an organ source. While this is obviously in violation of the rights of the kidnapped man, the fact that four other people and their families are made very happy by the arrangement makes it morally acceptable from the viewpoint of an act utilitarian.

By contrast, a rule utilitarian would look at the rule, rather than the act, that would apply to cutting up the kidnapped man for parts. Since the proposed applicable rule in this case—that one may kill a healthy man for his organs—if instituted widely would lead to particularly bad social consequences, a rule utilitarian would argue that

[20] As discussed in detail in Sect. 3.3, not all persons are human and not all humans are persons. Personhood requires that one (ordinarily) be in possession of continuous consciousness over time and that one be capable of mental representations concerning the world, as well as able to consider options, develop tactics and act on plans.

[21] It could be argued, of course, that our humanity is precisely situated in our ability and willingness to protect those who cannot fend for themselves.

we should in fact implement the opposite rule: "don't kidnap healthy people for their organs to transplant into sick people." Since negative utilitarianism requires us to act so as to produce the least amount of evil or harm for the greatest number of people, kidnapping a healthy man as a source of organs would obviously be immoral. In the case of Singer's preference utilitarianism, at issue is that all parties' preferences are met as much as possible. In the macabre example above, it is very likely that none of the parties involved would prefer to kidnap a healthy man as a source of organs.

It should be noted that, while utilitarianism has had a strong influence on the intellectual landscape of recent philosophical discourse and, in particular, in ethical theory, as in the example above, it sometimes falters badly when it is applied to questions of social or individual justice. Because classical utilitarianism seeks to maximize the total amount of a particular "utility" (like happiness or preferences) over an entire society or social group, it seeks whichever arrangement achieves maximum utility. But, as already emphasized, such an arrangement might be achieved by distributing benefits and burdens in a way that violates common notions of justice.

2.13 Critique of Utilitarianism

Utilitarianism has been attacked for reasons such as for violating common-sense notions of justice and fairness. Perhaps the most quoted example of how classical utilitarianism sometimes violates common-sense is the scenario where killing one individual would save the lives of many. As already noted, under the act utilitarian ethical model such action would be appropriate. As another example, such a situation arose in the 1968 movie "The Magus," where the mayor of a small Greek village under WW II German occupation is ordered by the Nazi Commandant to personally kill three Greek freedom fighters responsible for the death of German soldiers. If the mayor refused, the Germans would kill both the freedom fighters and all the villagers.

Another example: The use of slaves might greatly help maximize the *net* happiness in a society, but common-sense notions of justice almost always take slavery to be wrong (with apologies to both Aristotle and Thomas Jefferson, two great intellectuals who were unapologetic slave owners).[22]

Another serious criticism of utilitarianism is that under the goal of maximizing happiness or some other utility, the wishes and desires of sadists and perverts are lumped in with the wishes and desires of everyone else when an overall determination

[22]Thomas Jefferson appears to have felt that slavery was wrong but evidently never spoke against it publically, perhaps to avoid charges of hypocrisy. The January 1862 issue of *The Atlantic* makes passing reference to this issue. It can be read at http://www.theatlantic.com/past/politics/presiden/jeffslave.htm.

of utility is made. By espousing a system in which the satisfaction of *all* desires is to be maximized, utilitarianism can end up violating our intuitive precepts of natural justice.[23]

2.14 John Rawls

Such paradoxes as those identified above led the philosopher John Rawls and others to take the position that we must reject most forms of utilitarianism and instead develop a robust understanding of what is right and wrong as a basis for making ethical decisions. What is needed, Rawls argues, is a moral-political theory with justice-as-fairness at its core. Although a detailed explanation of his philosophy as set forth in his book, *A Theory of Justice* (Rawls 1971), is well beyond the scope of this discussion, in essence he argues that a fair and rational person operating behind a "veil of ignorance" would choose two general principles as the basis for social justice. The first principle is the *Principle of Equal Liberty*, where each person in a society would have an equal right to the most extensive liberties compatible with similar liberties for all. The second principle has two components. The first of these, called the *Difference Principle*, would require that any social and economic inequalities in a society should be the result of an arrangement that provides the greatest benefit to the least advantaged persons, while the second component, known as *Fair Equality of Opportunity*, holds that offices and positions must be open to everyone.

2.15 Virtue Ethics

Virtue ethics is yet another approach to handling ethical issues (Gardiner 2003; Begley 2005, 2008; Arries 2005). This third approach focuses on the character of the person (moral agent) handling the moral issue, not on any specific ethical rules (as the deontological approach emphasizes) or on the consequences of any ethical decision (as the utilitarian approach requires).

A virtue (Latin: virtus, Ancient Greek: ἀρετή "arete") is a characteristic or quality deemed to be morally good and thus suitable as a basis for moral excellence. Frequently identified virtues include courage, dignity, discipline, dutifulness, friendliness, industriousness, justice, mercy, modesty, prudence, temperance, tenacity, and truthfulness. The ancient Greek philosopher Aristotle emphasized the importance of developing a virtuous character in order for a person to flourish and achieve happiness *(eudaimonia)*. In his *Nicomachean Ethics*, he provides the reader with a list of

[23]Additionally, maximizing the happiness of all is not the same as maximizing 'the satisfaction of all desires', with the result that these two approaches could conceivably contradict each other. Also, a solution offered by one form of utilitarianism (e.g. one choosing the greatest good as the desired utility) would be expected to be different from a different approach, such as an approach based on preferences.

virtues similar to that offered above, and states that the individual who possesses a virtuous character must do the right thing, at the right time, in the right way, and for the right reason. As such, many philosophers regard Aristotle as the father of virtue ethics.

What differentiates virtue ethics from approaches such as consequentialism or deontology is the supremacy of virtue within the theory (Arries 2005; Gardiner 2003). While consequentialist philosophers might see virtues as characteristics that produce desirable outcomes and deontologists might see virtues as traits possessed by individuals who reliably complete their duties and responsibilities, virtue ethicists take a very different tactic. Simply put, virtue ethicists resist attempts to define virtue in terms of some other notion taken to be more fundamental and place virtue in its various forms as the fundamental element in virtue ethics.

Virtue ethics is not without its critics. One criticism is that, in principle, virtue ethics is often unable to provide specific guidance to direct moral action, as it is "agent-centered" rather than "act-centered" (Arries 2005; Begley 2005). Some virtue ethicists might respond to this concern by rejecting the assumption that virtue ethics ought to be in the business of providing an account of right action in the first place, but if this is the case there still remains a need for providing some sort of account to handle the practical ethical conundrums encountered in the real world. Proponents of virtue ethics who choose to retain the concept of right action additionally face the challenge of occasionally conflicting moral virtues, such as a conflict between justice and mercy that sometimes arises in situations where castigation and punishment must be meted out.

A practical clinical example may help differentiate virtue ethics from other ethical theories. Consider the question as to whether it is ever moral for a physician to lie to a patient "for their own good." For instance, an anesthesiologist might be asked by a patient about to undergo emergency surgery whether the surgeon about to do the case is highly qualified. Suppose that, in fact, the anesthesiologist does not hold the surgeon in high regard, since he seems to lose far more blood than other surgeons. However, he is concerned that the patient might refuse to undergo clinically necessary surgery should his reply not be favorable. What, then, should his reply be?

In this situation a deontologist would be expected to argue that lying to a patient is always wrong, regardless of any future "good" that would be expected to result, that the rightness or wrongness stems from the character of the act itself rather than the outcome of the act. A consequentialist would argue, however, that the morality of the act would depend on the expected outcome, and that lying to a patient might be appropriate if the result is that the patient's life is saved. Finally, a virtue ethicist would focus on what a virtuous physician would do in such a setting.

As can be seen from the above discussion, a number of moral theories are available to consider for adoption. What, then, are the characteristics of a "good" moral theory?

arly emphasize that a moral theory must necessitate respect for all sentient life s, not just humans.

ote also that some utilitarian philosophers such as Peter Singer (discussed ear- would likely take issue with any requirement that a moral theory be intuitive. er's moral positions are often seen to be counterintuitive and sometimes even gnant, especially in the matters of euthanasia and infanticide, although he makes ase lucidly in his many writings (Jamieson 1999). Singer and others in his camp e that reason and logic should trump moral intuition in making moral decisions. thought-provoking issue is considered in more detail in some of the following ons.

inally, one final qualification should be noted. It has been argued that rigid al theories are not useful because they do not provide guidance in individual or ptional instances where the situation demands interpretation of the theory or rule liams 1985). Ethical decision making is, more often than not, not merely a case pplying a rule. The realm of ethics demands interpretation and judgment on the of the involved parties. The challenge is to come up with a rule that is general igh to offer guidance across the board, while being 'flexible' enough to also v for application in cases that do not conform to what is generally applicable e population. Additionally, as soon as we venture into the field of ethics, we ssarily move in the world of meaning, where, in general, there are only opinions (often conflicting) interpretations of opinions.

7 Bernard Williams

is book, *Ethics and the Limits of Philosophy*, Williams (1985) argues that while osophical discourse on questions of an ethical nature can help us identify a variety ays to think about ethical problems, philosophy cannot justify why anyone *should* noral. Williams argues that any justification for morality, for example that moral s form the basis for a social contract that all reasonable people should agree to, or such rules maximize the overall well-being of society, or that moral rules result n our natural moral inclinations, will only appeal to people who already *want* to noral.

As an example, Kant famously justifies moral behavior on rational grounds, stating categorical imperatives[24] are duties that we must carry out to be rational actors. , argues Williams, Kant's line of argument fails in individuals who see no value in ng reason dictate their moral actions.[25] In dealing with utilitarian-based attempts

categorical imperative is an absolute, unconditional moral requirement that applies to all ımstances. It is distinguished from a *hypothetical imperative*, where certain ends apply (e.g., if sh to tame my hunger, I must eat a meal).

r example, many religious traditions have rules of behavior that are revealed via sacred scripture er than being the result of rational discourse. Wrote Martin Luther: "Reason is the greatest enemy faith has; it never comes to the aid of spiritual things, but more frequently than not struggles

2.16 Characteristics of a Good Moral Theory

Any good moral theory has a set of traits that characterizes the tl sophically sound. These characteristics are needed to avoid a num flaws that might otherwise occur. These include: bias, cultural i ideology, prejudice, racism, sexism and other defects in logic ar

I would also argue that the following traits are valuable (essential) characteristics of any moral theory. First, a moral consistent—i.e., be free of internal contradictions and yieldin similar settings. Suppose, for example, that a soldier believes disobey his commandant (having taken an oath of obedience) an it is wrong to disobey the recognized rules of war (having taken a effect). Suppose further that one day the commandant insists that t prisoner in an attempt to obtain intelligence. The situation leads t among moral principles: the soldier can either obey his comma the rules of war, but not both.

A second trait is that the theory should be universal—i.e., if to one individual, then it should apply to all individuals. For exa of the torture example above, the universality of "just war" moral combatants holds that American soldiers (as an example) are not rules of war with respect to the torture of prisoners.

Finally, a moral theory should also be intuitive—i.e., the theo fit our moral intuition. In the requirements that a moral theory b and universal, it will be immediately understood that an inconsist that applied only to some people but not others would be undesirab less obvious that a moral theory should fit our moral intuition, sinc may have strong cultural influences. On the other hand, a moral th intuition is more likely to be understood and followed.

Additionally, ethical action is often counter-intuitive; some ethic go against one's intuition, but makes sense when rationally consider my intuition tells me to help beggars by giving them money; howeve remain beggars because the money received supports an addictio them disempowered. Therefore, rationally speaking, it sometimes n to merely give money to beggars, but to seek other ways to help ther

To this list one might add other possible desirable characteristics, su the theory to be understandable by non-philosophers (certainly a requ practical theory), or the need for the theory not to be based on any reli (although arguably this is already covered by the second requireme might also add the requirements of any moral theory being time-inv principles hold true over time) and trans-cultural (that the principl cultures), but these requirements are arguably also covered by the secon above.

One might also axiomatically take the position that any moral theor all forms of human life, no matter how degraded, while animal rights ad to justify morality, Williams doubts whether the analytic impartiality utilitarianism requires is actually possible, even if agreement might be found concerning the best variant of utilitarian analysis to invoke. In sum, Williams fears that any attempt to justify morality in a way that does not presuppose an existing commitment to morality will fail. Instead of providing a system of moral philosophy, Williams steadfastly maintained an overarching skepticism about attempts to create a solid foundation for moral philosophy. Somewhat similar views have been offered in his book *Shame and Necessity* (Williams 1993), where he argued that moral theories are insufficient to deal with real life complexities, particularly given the deep-seated pluralism encountered in contemporary societies. Our ethical life, Williams contends, is simply too muddled to be captured by any systematic moral theory.

2.18 Reason Vs. Moral Intuition

Earlier, the characteristics of a "good" moral theory were discussed. Of this list of characteristics, one characteristic in particular remains the subject of considerable debate and discussion: whether or not a moral theory should always fit our moral intuition. It is on this point that an interesting controversy has arisen that has sharply divided the academic bioethics community. Distilled to its essence, the debate centers on whether moral intuition should take precedence over reason and logic in making ethical decisions. This issue is discussed next.

Consider the following example. Should the lives of severely deformed neonates ever be actively terminated? Some moral philosophers such as Peter Singer regard this as sometimes acceptable, even desirable, arguing that this follows as a consequence of utilitarian ethical principles. Others, such as Leon Kass, argue that the natural repugnance of this act should serve as a clear moral guide against such actions.

Kass, in his famous essay, the "The Wisdom of Repugnance" (Kass 1997) argues that we should avoid interventions such as genetic human enhancement based on the simple notion of repugnance. That is, Kass argues that notions that are revolting to our sensibilities—for instance, the idea of making human-animal hybrids—should be viewed as being unethical on the basis of the revulsion they generate (rather than, as Singer has argued and most transhumanists would be expected to argue, on the basis of the moral consequences they entail.) Kass and his philosophy is discussed in more detail below and in Sect. 7.4.

---

against the divine Word, treating with contempt all that emanates from God." (www.godlessgeeks.com/LINKS/Quotes.htm).

2.19 Sanctity of Human Life

Leon Kass and his bioconservative philosophical schoolmates view Singer's philosophical positions and many of those of the transhumanist movement as bizarre and absurd, since they are so profoundly counterintuitive. They argue that the notion of "sanctity of human life", even if only a mere intuitive notion, should triumph over all other philosophical positions in matters regarding human life. Other bioconservatives in Kass's fold argue that Singer's "absurd" conclusions follow necessarily as logical outcomes of a defective moral theory, preference utilitarianism. They argue that if there is a problem with Singer's conclusions, it is entirely because of flaws inherent in the moral theory of preference utilitarianism, and not from flaws in Singer's logic. Deploying a variation of the ancient "*reductio ad absurdum*" argument (in which a proposition is disproven by following its implications to a logical but absurd consequence), they argue that since Singer's impeccable logic yields completely ridiculous results, the philosophical foundations of preference utilitarianism are thereby proven to seriously flawed.

2.20 Further Concerns About Singer

Concerns about Singer's views on utilitarianism and personhood have arisen from a number of conservative philosophers other than Kass. Charles Weijer, a bioethicist at Dalhousie University in Halifax, Canada, elaborates on some of Singer's radical positions (Weijer 2000): "Singer variously concludes that experimenting on a human embryo is preferable to doing so on a mouse; that chimpanzees are properly called "people," but humans with profound cognitive impairments are not; and the heart of a handicapped newborn human might be legitimately excised to save a baboon in need of a new heart." Philosophers such as Kass and Weijer see Singer as being irresponsible in letting logic dominate over intuition. For instance, Charles Weijer later states: "Doing ethics responsibly involves more than logical reasoning alone. Moral intuition acts as an important check on ethical reasoning, telling us that at times it is the theory, not our actions, that must be changed."

Similar concerns have been offered by Donald Demarco in his article, "Peter Singer: Architect of the Culture of Death" (DeMarco 2003), where he writes:

> Professor Singer underscores the importance of reason, broadmindedness, and compassion. But his emphasis on reason displaces human feelings. His advocacy of broadmindedness causes him to lose sight of the distinctiveness of the human being (he does not object to sexual "relationships" between humans and non-human animals). And his sensitivity for compassion is exercised at the expense of failing to understand how suffering can have personal meaning. In the end, his philosophy is one-sided and distorted. It plays into the Culture of Death because it distrusts the province of the heart, fails to discern the true dignity of the human person, and elevates the killing of innocent human beings — young and old — to the level of a social therapeutic.

2.21 Logic in Philosophical Discourse

As the above discussion makes clear, Kass, Weijer, Demarco and others would have us place human feelings and moral intuition over logic and rational discourse, at least in some settings. However, any philosophical position emphasizing the role of moral intuition over logic and rationality has its own potential flaws. From the time of the ancient Greeks, philosophers have repeatedly emphasized the importance of logic in philosophical discourse. Aristotle held that we can use reason to determine the best way to achieve the highest moral good. More recently, Kurt Baier in his book, *The Moral Point of View* (Baier 1958) argued that moral decision making involves a search for the best reasons for or against a particular course of action. He notes that moral reasoning parallels legal and scientific reasoning whereby one collects relevant facts and weighs arguments on all sides of the issue. Only then is a judgment made. By contrast, appeal to moral intuition bypasses this process. Thus, an appeal to moral intuition raises a number of rather difficult questions. Why suspend logic merely because the results of its application are unappealing? And how does "moral intuition" differ from mere "gut feeling"?

Clearly, the moral intuition approach has potential dangers. Ethical issues become purely social-cultural considerations rather than based on logic and evidence, with the consequence that cultural relativity predominates moral thought. (Cultural ethical relativism teaches that morality is relative to a given culture and that it is wrong for individuals in one culture to judge the moral practices of another culture.) Furthermore, ethical relativism violates the principle of universality presented earlier. Finally, as discussed in Chap. 7, the moral intuition approach sometimes suffers from a fallacious appeal to emotion: "It's disgusting, so it has to be wrong!" The ultimate underlying premise in the moral intuition position seems to be that emotional responses should trump the rule of reason.

2.22 An Axiomatic Approach

Which approach, then, is the most philosophically solid? As noted, earlier, Singer's camp suffers from a problem common to most forms of utilitarianism—that of producing counterintuitive moral proclamations—while followers of Kass, holding that we should respect moral intuitions about the special value of being human, cannot identify reasons to support such notions. Since intuition is often subject to serious cultural and cognitive biases, this approach is problematic as well.

Because the intuition approach to ethical quandaries allows one to bypass formal, rational philosophical analysis in favor of what ultimately amounts to a "gut feeling", one should have serious reservations about this approach. After all, if philosophy is about anything, it is about the use of logic, analysis and reason. But one should also be somewhat uncomfortable with Singer's utilitarianism. For example, how does one ultimately decide which particular flavor of utilitarianism to choose as a philosophical foundation? Also, why would Singer have us pick preference utilitarianism over earlier forms, such as act utilitarianism? Since all forms of utilitarianism have philosophical flaws, a sound means of selecting among them is needed. Unfortunately, no philosopher seems to have successfully tackled this difficult problem.

All mathematicians recognize that all systems of mathematics, such as Euclidean geometry, start from an initial set of axioms. Axioms are starting principles that cannot be proven to be true; their truth is so immediately evident that no reasoning or demonstration can make it plainer. Similar to Euclid's axioms, in philosophy there are some principles that one cannot imagine to be otherwise. An example is Aristotle's law of contradiction: A cannot be B and not be B at the same time and in the same manner. An even better-known example is Descarte's *Cogito.*[26] And just as in mathematics, philosophical axioms cannot be proven—they are either accepted as self-evident or rejected.

However, one guiding principle that has not seen universal acceptance concerns the sanctity of human life. While some moral philosophers argue that the maintenance of human life by ordinary means, even in severely degraded forms, should be an axiomatic moral imperative, others would allow actions such as suicide and euthanasia, at least under some circumstances. It is on this principle that a great divide in moral philosophy can be seen to exist.

As just emphasized, all proposed solutions to bioethical dilemmas are entirely predicated on which starting philosophical principles or axioms one chooses. Unfortunately, individuals seeking to analyze bioethical dilemmas appear to be left utterly to their own devices to decide upon a set of starting philosophical principles; they are expected to draw on their own moral or philosophical instincts as a guide to making this choice. Of course, relying on one's instincts would be expected to produce a number of incompatible ethical approaches, just as political opinion inevitably varies among individuals asked to provide a political judgment. I suppose this is why philosophers so frequently disagree. Kimberly (2002) has summarized the disagreement (in the context of human cloning) as follows:

> Kass proposes that society yield to repugnance as an ethically relevant factor, since the elicitation of such repugnance signifies the defilement of human nature. However, such an application of repulsion and offense to human nature leads to improbable conclusions and internal contradictions that soundly repudiate the acceptance of such a principle. Thus, rather than pre-analytically yielding to visceral emotions and passions, individual rational agents–and the public generally–ought to thoroughly analyze all of the relevant factors surrounding cloning before, and perhaps instead of, rejecting the technology simply because it elicits repugnance.

[26] *Cogito ergo sum* is a Latin proposition offered by René Descartes in his *Discourse on Method*. Usually translated into English as "I think, therefore I am" it became a founding pillar of Western philosophy, purporting to form a secure foundation for knowledge in the face of radical doubt (wherein one refuses to accept anything as true unless it appears to be "clearly and distinctly" true). Descartes argued that while other kinds of knowledge could actually be a figment of one's imagination, a deception, or a mistake, the very act of questioning one's own existence necessarily confirms the reality of one's own mind.

Ultimately, then, it seems how one views Singer and Kass and their respective camps depends on which philosophical foundations one eventually (but perhaps arbitrarily) chooses as a starting point for moral philosophy. The moral rationalism approach is based heavily on reason and logic, following in the philosophical footsteps of Aristotle. By contrast, the moral intuition approach relies on instinctive notions as to how things should be, bypassing traditional philosophical methods of analysis for an approach that relies on gut feeling and emotion. In the end, while both approaches can be shown to have noteworthy philosophical difficulties, my opinion is that Singer's moral rationalist position can be seen to be philosophically more robust in the sense that it utilizes traditional philosophical methods of analysis rather than relying on intuition and emotion.

2.23 Consent

Another issue of central importance in bioethics is consent. As noted earlier, the historical origins of the principle of autonomy stems in part from a need to develop a philosophical framework for ethically managing randomized controlled trials (where patients are administered one of a number of forms of medical treatment on the basis of chance alone). Additionally, it originated as a response to clear ethical abuses that have occurred in the past. Finally, a number of legal decisions have reinforced the notion of patient autonomy.[27]

Informed consent arises in two principal contexts; clinical and experimental, although overlap between the two does occur. In the clinical context, patients are provided with information pertaining to the risks, benefits, and the alternatives to proposed medical interventions such as surgical procedures. Usually, but not always, the process includes the use of a paper consent form that is signed, dated and witnessed.

In the experimental context, research subjects (who in medical research studies are often also patients) agree to participate in an approved experimental protocol that may not have the benefit of the experimental subject as a primary or even as a likely outcome (e.g., testing the safety of a new drug on healthy human volunteers).

In either case, establishing informed consent from the individual is the central ethical concern, and both require that the individual is capable of understanding the issues involved and (in the case of research studies) additionally require that special safeguards be employed if the individual is a member of a "vulnerable" population, such as children, prison inmates, or pregnant women.

Consent can be implied or explicit, the latter involving a more or less formal process carried out verbally with or without explicit documentation in writing. Implied consent exists (for instance) when one unhesitatingly rolls up one's shirt sleeve in

[27] *Schloendorff vs. Society of New York Hospital* and *Malette vs. Shulman* are two notable examples. Both these important legal cases are discussed in Sect. 2.29.

preparation for a blood test or when an unconscious patient is taken to hospital for emergency treatment of life-threatening injuries.

Explicit consent is usually sought prior to clinical interventions that entail some risk of harm or have the potential to cause a substantial degree of pain or discomfort. Some institutions have a policy that a consent form must be signed by the patient prior to surgical procedures but may not have a similar requirement for other risky interventions such as blood transfusion, central line insertion, or lumbar puncture. If a signed and witnessed consent form is not required, and the intervention involves a non-trivial risk, clinicians are usually advised to write a detailed note in the patient's medical record to establish that the consent process has taken place, with specific mention of risks, benefits, and alternatives.

As emphasized earlier, central to informed consent is an explanation of the risks, benefits, and alternatives associated with any proposed clinical or experimental intervention. However, in some cases, such data may not be fully available, while, in some other cases, the amount of information available may be so large that it has the potential to overwhelm even well-informed and experienced patients. In addition, special problems may occur in providing risk information to patients. For instance, the only risk information available for an intervention may be that for the medical community at large and may not be specific to a particular institution or a particular clinician.

In some situations, such as complex cancer treatments, the issues involved may be so multifaceted that, at times, only highly-trained individuals can fully understand the issues involved. This is particularly true in controversial areas of medicine where competing viewpoints may arise as a result of methodological and even cultural differences. As a result, providing full information may potentially require special methods of patient education or may be so complex that it is beyond the cognitive capabilities of some patients.

A final issue concerns how much information to provide to patients. Some clinicians and bioethicists suggest that patients should be told of common risks with low morbidity as well as rare risks with a high associated morbidity, but need not necessarily be provided with an exhaustive list of all possible risks regardless of their likelihood or their severity. Of interest, this is precisely the position taken by some legal authorities (vide infra).

Finally, it may be helpful to distinguish between the risks of a medical intervention and any uncertainties that may be associated with the procedure. An example will help explain the distinction. In some cases, the risks of a common procedure have been well-established by actuarial means, such as the risk of a stroke or death following coronary artery bypass grafting surgery. But when an individual patient deviates significantly from the sample population upon which the actuarial data is based (for example, being extremely elderly, or having a rare co-morbidity that would be expected to predispose to perioperative complications) this risk data is no longer accurately representative of the true risks involved, and additional uncertainties are necessarily introduced into the discussion.

2.24 Medical Futility

A number of ethicists argue that in cases where there is no hope for improvement for a medical condition such as terminal cancer or advanced AIDS that no course of treatment is called for other than "comfort measures" (Weijer et al. 1998; Gampel 2006; Burns and Truog 2007; Mickelsen et al. 2013). That is, to apply treatments with zero chance of efficacy is wasteful of valuable resources and only serves to provide false hope. Of course, not every clinician or patient will agree as to when no further benefit to an intervention is to be expected. Still, there are situations where agreement amongst clinicians is almost universal. An example is the patient who properly meets the criteria for brain death. Such patients have zero chance of recovery and are often used as a source of organs for transplantation. Family members, however, seeing a rhythm on the electrocardiogram display and feeling a pulse at the wrist, and hoping for a miracle, sometimes request continuing care in an Intensive Care Unit (ICU) even though such care is futile. This creates an ethical dilemma for some clinicians who feel that it is wrong to provide medical care to a (brain) dead person (Luce 1995; Burt 2002; Bernat 2005).

A key case in the debate on medical futility was Baby K. Baby K was born in an anencephalic state on October 13, 1992, at Fairfax Hospital in Fairfax, Virginia. She was born missing almost all of her brain; in fact, all that remained was the "brainstem", that primitive part of the brain responsible (in part) for autonomic and regulatory function, such as the control of respiration, the heart beat and blood pressure. In almost all cases anencephalic infants are not aggressively resuscitated since there is zero chance of the infant ever achieving a conscious existence. Instead, the usual clinical practice is to offer hydration, nutrition and comfort measures and "let nature take its course." Artificial ventilation, surgery (to fix any co-existing congenital defects), and drug therapy (such as antibiotics) are usually regarded as being pointless. Some clinicians see no point in even providing nutrition and hydration, arguing that withdrawal of nutrition and hydration is morally and clinically appropriate in such cases.

One should understand that anencephalic babies are technically not brain dead, as they usually have intact brainstem reflexes. Yet there is strong clinical consensus that valiant efforts should not be employed to keep these infants alive. In fact, anencephaly and brain death may be the only two clinical situations that virtually all knowledgeable clinicians agree are futile to treat (except possibly to the extent necessary to allow organ harvesting).

Occasionally parents want clinicians to use all available means to keep anencephalic infants alive as long as possible. However, in most of these cases parents eventually come to realize that there is no possibility of a good outcome from such efforts and end up agreeing with the clinical team. Despite this, in the case of Baby K., Ms. H., the mother, wanted the hospital to continue with advanced supportive care (primarily ventilatory support) against the wishes of the clinical team, and sought legal support for her position. Ms. H. knew of her baby's condition from the second trimester of her pregnancy, but, motivated by a strong religious conviction that "all life is precious" and that God alone should decide how long the baby would live, she remained adamant that Baby K. be kept alive as long as possible.

The hospital's position was that such care would be futile. At the trial,[28] expert testimony was given to demonstrate that provision of ventilator support for anencephalic infants goes beyond the accepted standard of care. The legal team for Baby K's mother adhered to a religious sanctity-of-life principle as the basis for their case. In the end, in a particularly controversial decision, the U.S. District Court ruled that the hospital caring for Baby K must put her on a mechanical ventilator whenever she had trouble breathing. In particular, the court interpreted the Emergency Medical Treatment and Active Labor Act (EMTALA) to require continued ventilation for the infant. The wording of this act requires that patients who present with a medical emergency must get "such treatment as may be required to stabilize the medical condition" before the patient is transferred to another facility. The court took the position that "it is beyond the limits of our judicial function to address the moral or ethical propriety of providing emergency stabilizing medical treatment to anencephalic infants. We are bound to interpret federal statutes in accordance with their plain language…" As a result of the decision, Baby K was kept alive much longer than most anencephalic babies, living to age 2 ½.

2.25 Doctrine of Double Effect

The "doctrine of double effect" (Saini 1999; McIntyre 2004) is a principle of ethics potentially applicable in many ethical dilemmas. First espoused by St. Thomas Aquinas in his *Summa Theologica,* the doctrine states that it is sometimes permissible to cause a harmful side effect in bringing about a good end result even though it would not be ethical to cause such a harm directly. According to the *Stanford Encyclopedia of Philosophy* the doctrine "is often invoked to explain the permissibility of an action that causes a serious harm, such as the death of a human being, as a side effect of promoting some good end". [29]

As a case in point, the *Stanford Encyclopedia of Philosophy* goes on to provide the following example: "A doctor who intends to hasten the death of a terminally ill patient by injecting a large dose of morphine would act impermissibly because he intends to bring about the patient's death. However, a doctor who intended to relieve the patient's pain with that same dose and merely foresaw the hastening of the patient's death would act permissibly."

[28] *Matter of Baby K*. 16 F.3d 590 (4th Cir. 1994), n. 9 at 598.

[29] http://plato.stanford.edu/entries/double-effect/.

2.26 Therapeutic Privilege

The term "therapeutic privilege" refers to the withholding of material information by the clinician during the "consent" process in the belief that the disclosure of such information would lead to the psychological harm or unnecessary suffering for the patient. That is, therapeutic privilege is invoked as an exception to the doctrine of informed consent.

The philosophical argument behind "therapeutic privilege" is that beneficence should take precedence over autonomy in cases where a conflict between these two principles exists. Historically, therapeutic privilege was invoked by some doctors when they withheld the diagnosis of a terminal condition from patients whom they expected would not be able to handle the information, and would consequently lose an interest in living. In many cases, such information was withheld at the specific request of family members (Elwyn et al. 2002).

Another example concerns patients born with ambiguous genitalia. Some past practices involved the use of deception and/or incomplete communication of facts about the infant's condition and early surgical intervention to make a "definitive" sex and gender assignment (Côté 2000).

Today, with the collapse of paternalism in modern clinical care, therapeutic privilege is invoked only rarely, if at all. This is so both as a result of medico-legal prudence (vide infra) and in response to a cultural shift that emphasizes individual autonomy more than in previous times. Furthermore, some clinicians argue that, if a person feels that providing a patient with relevant clinical information may be so upsetting to the patient as to cause extreme psychological harm, the patient should be assessed for mental competency.

Still, therapeutic privilege has its cautious supporters. For instance, (Etchells et al. 1996), discussing the Canadian situation, make the case that therapeutic privilege may sometimes be appropriate. They write:

> The need for sensitivity to cultural norms may potentially support the exercise of therapeutic privilege. In some cultures therapeutic privilege is widely invoked, and it is unclear whether patients from these cultures should always be subjected to Western standards of consent. However, given the legal status of therapeutic privilege in Canada, clinicians should avoid invoking therapeutic privilege. It is better for the clinician to offer information and allow the patient to refuse or accept further disclosure.

2.27 The Permissible Death Problem

Few legal drugs are used as ubiquitously as Tylenol, the popular nonprescription analgesic known by the scientific names of acetaminophen (and paracetamol). Although the use of acetaminophen is advocated for a number of mild-to-moderate pain conditions such as headaches and arthritis, it is less well-known that in large doses acetaminophen can be lethal. In fact, acetaminophen liver toxicity, often from an intentional overdose, causes more than 450 deaths annually in the United States, and this number appears to be on the rise (Bridger et al. 1998; James et al. 2003; Larson 2007). While there is little doubt that this number could be substantially reduced by restricting access to the drug, such as removing its nonprescription status, or even removing the drug from the market entirely (as happened with the painkiller Vioxx), the fact is that regulatory authorities like the U.S. Food and Drug Administration (FDA) must view this number of deaths as being acceptable in light of the enormous overall clinical benefits the drug provides. This example illustrates the concept of "permissible deaths", a thorny ethical issue that regulators and policy makers must deal with in a great many settings. In the context of this book, this difficult problem arises in the context of risk versus benefit in a variety of potential transhumanist interventions (e.g., genetic engineering methods, pharmaceutical enhancement technologies, neural augmentation via surgical implants) that may become widely available in the future.

The problem of "permissible deaths" is hardly unique to the regulation of drugs; there are many other instances where this nasty issue shows up. For example, generals conducting military campaigns must decide how many combatant deaths on either side are tolerable, as well as decide how many unavoidable innocent civilian deaths are acceptable ("collateral damage") (Alexander 2001; Arya 2007).

Similarly, safety engineers must strike a balance between the cost of a safety feature and the number of lives saved, since, for instance, relatively few people would be willing to buy a car costing $300,000 no matter how safe (Viamonte et al. 2006). Likewise, adding numerous extra safety features to nuclear power plants, to commercial aircraft or to invasive medical equipment could conceivably make these products too expensive to be affordable.

This issue even comes up in surgery. In high-risk procedures such as cardiac surgery, how many deaths are acceptable? In 1998 the British General Medical Council, the regulatory agency that monitors British doctors, charged that two heart surgeons under their scrutiny were guilty of operating on children despite knowing that their fatality rates were unacceptable (Treasure 1998; Bolsin 2002). This naturally raises the issue as to what an acceptable death rate might be and how such a rate should be determined. (In some US states like New York where the fatality rates for all heart surgeons are publicly available, one unintended consequence has been for heart surgeons to simply refuse to take on very high-risk cases for fear of adversely affecting their "batting average".)

Another medical situation concerns permissible deaths related to the use of medical equipment. For instance, patient-controlled analgesia (PCA) machines that allow patients to self-administer morphine after painful surgical procedures are inherently risky. While potential benefits of this technology include superior pain control, automatic electronic documentation and improved utilization of nursing resources, unfortunately, unanticipated flaws in the design of these machines can sometimes lead to adverse drug events such as overdoses, and even death (Lin et al. 2001; Vicente et al. 2003). One particularly notorious unit is the Abbott Lifecare 4100 PCA Plus II machine. In 1997 three deaths that occurred while patients were connected to this device were documented. Investigations revealed that part of the problem was an unfriendly user interface that made user errors more likely. Unfortunately, despite being amply notified of this problem, the manufacturer was unwilling to upgrade the unit to a safer design, claiming that there was no problem with the unit in the first place that could not be handled with proper user training. In the end no design change was ever implemented and the unit remained in clinical use until recently (Doyle 2007).

Clearly, complex technologies like automobiles and nuclear power plants offer personal and social benefits at a price that necessarily produces occasional injury and death. Still, when government regulators license drugs or medical devices, they implicitly require that the perceived benefits exceed the perceived risks. In the case of drug products, when this relationship is no longer obvious, the drug may be withdrawn (as happened to the drug Vioxx in the USA) or its indications restricted (as happened to the drug Avandia in the USA). In the case of medical devices like PCA machines, when preventable deaths or injury have occurred regulatory agencies may require that the device be recalled from clinical service so that safety upgrades can be instituted. But, as in the case of the Abbott PCA machine, this is not always the case.

Failure to mandate a recall of medical devices that harm patients may occur for several reasons. First, the remedy may be so expensive as to be impractical. Second, the medical device may involve an old design that is approaching the end of its life cycle anyway. Third, regulatory agencies with limited resources must prioritize their goals, with the result that medical devices that injure or kill only a small handful of people may not get the regulatory attention that victims and their families would otherwise like. In such cases, legal remedies may be the only option available.

2.28 Legal Perspectives

A number of bioethical issues have emerged as legal cases. Here we offer a small sampling of such cases from Australia, Canada, England and the USA. They are presented here with the goal of presenting some important real-world bioethical issues from a legal standpoint. In particular, these cases show how the concept of patient autonomy and the related notions of informed consent and individual liberty have become firmly established in modern bioethical discourse, reflecting a change in mindset from paternalistic attitudes of earlier eras.

Schloendorff Vs. Society of New York Hospital

In the USA, the legal case that definitively established the right of competent adults to refuse medical treatment was tried in 1914 in a case known as Schloendorff vs. Society of New York Hospital.[30] In this case, the plaintiff agreed to undergo a gynecologic examination under general anesthesia, but she explicitly refused consent for any surgical intervention (Friedlander 1995). However, at the time of the procedure,

[30] Schloendorff vs. Society of New York Hospital, 105 N.E. 92,1914.

surgically correctable pathology was identified, and the surgeon decided to correct the problem despite instructions to the contrary.

Unfortunately for all parties, serious unexpected complications followed the surgery, and, as a result, a lawsuit was launched. The litigation was resolved in favor of the plaintiff. In his opinion, the presiding judge wrote, "Every human being of adult years and sound mind has a right to determine what shall be done with his own body" (Sreenivasan 2003). (Given that the plaintiff was a woman, the gender in the wording of the judge's opinion can only be viewed with dismay!)

Despite this favorable legal ruling, in the end, the patient actually lost her case because the hospital was a charitable institution and consequently was immune from liability under the laws of the time. Still, this case firmly established the notion of informed consent and of the right of a competent adult patient to choose or refuse medical treatment. Since the time of this landmark case, a large number of American legal decisions have reinforced the right of a competent adult to choose his or her treatment, even when the decision is not clinically sound.

Malette Vs. Shulman

The Jehovah's Witnesses are a Christian religion whose followers believe the Bible to be the true and literal word of God. Witnesses view Jehovah as the ultimate moral authority and do not salute flags, join service organizations, enlist in the military, vote in public elections, or take an interest in civil government.

Traditionally, orthodox Jehovah's Witnesses will not accept homologous or autologous whole blood, packed red blood cells, plasma, platelets and white blood cells, even when "clinically necessary" (Letsoalo 1998; Ridley 1999). The determination that blood transfusions were a violation of God's law was made in 1945 and is primarily based on three biblical passages (Benson 1989):

> Every moving animal that is alive may serve as food for you. As in the case of green vegetation, I do give it all to you. Only flesh with its soul – its blood – you must not eat. (Genesis 9:3, 4)

> As for any man of the house of Israel or some alien resident who is residing as an alien in their midst who eats any sort of blood, I shall certainly set my face against the soul that is eating the blood and I shall indeed cut him off from his people. (Leviticus 17:10–16)

> The Holy Spirit and we ourselves have favored adding no further burden to you, except these necessary things, to keep abstaining from things sacrificed to idols and from blood and from things strangled and from fornication. If you keep yourselves from these things you will prosper. (Acts 15:28, 29)

This prohibition can result in a challenging dilemma for physicians because a routine, safe, and potentially life-saving medical intervention is unacceptable to the patient (Rothenberg 1990; Mann et al. 1992; Benson 1989). Anesthesiologists are particularly affected because they are almost always responsible for intraoperative transfusion management. This sometimes puts clinicians in very difficult ethical situations (Harnett et al. 2000; Baldry et al. 2000).

In Canada, an important legal precedent regarding patient consent in this respect was the case of Malette vs. Shulman (Sneiderman 1991) . In this landmark case, the court established that emergency treatment should not be administered without patient consent if there is substantial reason to believe that the patient would refuse the treatment if he or she were able.

The specifics of the case are as follows. Dr. Shulman was an emergency room doctor caring for a woman who was unconscious as a result of hypovolemic shock from blood loss following a serious motor vehicle accident. After a quick clinical assessment, Dr. Shulman administered a medically necessary blood transfusion to the patient, saving her life as a result. Unfortunately for Dr. Shulman, the patient carried a wallet card indicating that she was a Jehovah's Witness and did not want to receive a blood transfusion under any circumstances. Also, complicating this situation was the fact that the card was written in the French language, and was signed but undated and unwitnessed, thus casting some doubt on the legal significance of the card, at least in the mind of Dr. Shulman. However, in the end, Dr. Shulman weighed the pros and cons of transfusing and decided to go for life.

Although the blood transfusions Dr. Shulman administered were responsible for saving the patient from death, the patient sued. To the surprise of the Canadian medical community, the court found that Dr. Shulman was liable for battery. In his decision, the judge wrote: "To transfuse a Jehovah's Witness in the face of her explicit instructions to the contrary would, in my opinion, violate her right to control her own body and show disrespect for the religious values by which she has chosen to live her life" (Sneiderman 1991).

Rogers Vs. Whitaker

An important Australian case known as Rogers vs. Whitaker involved the issue whether a doctor has a duty to warn patients of any significant risk involved in a proposed procedure (Chalmers and Schwartz 1993). In this case, the plaintiff lost her sight after unsuccessful surgery in her right eye, which eventually led to "sympathetic" blindness in her left eye. This was a rarely encountered complication that occurs in approximately 1 out of 14,000 cases. The plaintiff argued that, while there was no question that the surgery had been performed with skill, the surgeon's failure to warn that blindness in the good eye could develop constituted negligence. The defense position was that the doctor acted within the purview of common practice, a position that the court ultimately rejected.

Hills Vs. Potter

The courts have not universally supported a position such as that of Rogers vs. Whitaker. For example, in the British case of Hills vs. Potter,[31] a patient who developed paralysis following elective neck surgery sued, arguing that the surgeon failed to provide information necessary to make an informed decision whether to undergo the surgery. However, the court dismissed her claim, rejecting the doctrine of informed consent commonly prevailing in American and Canadian law, under which a physician has a duty to disclose all material risks which patient might view as being

[31] Great Britain. England. Queen's Bench Division. Hills vs. Potter. All Engl Law Rep. 1983; 3:716–29.

important. Legal decisions such as Hills vs. Potter serve to illustrate how courts in various nations, on the basis of differing legal precedents and other (e.g., cultural) factors, may take very different perspectives on similar issues.

2.29 Conclusion

This chapter was written to provide a background for issues to be discussed in subsequent chapters. These include the criteria for humanness and personhood, the moral standing of human and nonhuman sentient beings, the morality of human enhancement via technical and pharmacological means, and related concerns. In particular, in Chaps. 5 and 6 it is shown that the line between life and death is at times more apparent than real, a fact that—as well shall see later—raises important challenges to moral philosophers and medical practioners alike.

Chapter 3
Humans, Transhumans and Humanoids

Man is an artifact designed for space travel. He is not designed to remain in his present biologic state any more than a tadpole is designed to remain a tadpole.

William S. Burroughs (http://quotes.dictionary.com/Man_is_an_artifact_designed_for_space_travel)

Abstract This chapter introduces the reader to the field of transhumanism and some of the most important philosophers contributing to this recent movement. Related philosophical issues such as moral standing and personhood theory are presented. The possibilities of genetic engineering as a means to fundamentally alter the nature of humankind are discussed, along with the concepts of animal-human hybrids, computer-human hybrids, and fully artificial beings. These ideas are presented to support the premise that there is no unique and unalterable biological form that constitutes humanity, and that our concept of what it means to be human will be forced to evolve in the face of new biotechnologies such as genetic engineering, molecular biology, and nanotechnology, all of which have the potential to alter our nature and character in ethically profound ways.

3.1 Introduction

Following on the discussion of ethics and ethical theory introduced in the previous chapter, which summarized and critiqued the various approaches to tacking problems in moral philosophy, this chapter introduces the transhumanist movement and some of the most important philosophers contributing to this movement. Related philosophical issues such as moral standing and personhood are presented. The possibilities of genetic engineering as a means to fundamentally alter the nature of humankind are discussed, along with the concepts of animal-human hybrids, computer-human hybrids, and fully artificial beings. These ideas are presented to support my contention that there is no unique and unalterable biological form that constitutes humanity, and

D. J. Doyle, *What Does it Mean to be Human? Life, Death, Personhood and the Transhumanist Movement*, Anticipation Science 3,
https://doi.org/10.1007/978-3-319-94950-5_3

that our concept of what it means to be human will be forced to evolve in the face of brave, new biotechnologies such as genetic engineering, molecular biology, and nanotechnology, all of which have the potential to alter our nature and character in ethically profound ways.[1] Readers interested in learning about recent books dealing with transhumanism are referred to the Appendix at the end of this work.

3.2 Moral Standing

The notion of moral standing is a central concept in ethical discourse, particularly as it relates to the treatment of nonhuman individuals. This notion is presented and developed here in order to show how, just as the moral treatment of animals came to be an important philosophical topic in earlier decades, it arises once again in the discussion as to what constitutes the moral treatment of entities such as animal/human hybrids, computer/human hybrids, and other beings with a human component (sometimes known as "parahumans").

The "moral standing" of a being determines the degree to which its welfare and well-being must be given ethical consideration. That is, a being merits moral standing if we believe that it makes a difference, morally speaking, how that being is treated, and that we should take into account that individual's interests for the individual's own sake and not merely for our benefit or the benefit of someone else. Consider the following helpful example from the Markkula Center for Applied Ethics[2]:

> Take, for example, a doctor who attends to the physical welfare of her patients and believes that it would be morally wrong to mistreat them. Suppose that she believes this, not because of any benefits she will derive from taking good care of them nor because she is afraid of being sued, but only because she has a genuine concern for her patients' well-being. Her patients have moral standing for her. On the other hand, take a farmer who looks after the welfare of his cows and who also believes that it would be morally wrong to mistreat them. But suppose he believes this only because mistreating them would decrease their milk production and their milk is an essential source of nourishment and income for his family. Although this farmer considers his cows' welfare, he does so only for the sake of his family and not for the sake of the cows themselves. For the farmer, the cows have no moral standing.

A great philosophical question that continues to this day asks the question whether any beings other than humans merit moral standing. One commonly held position is that only beings with the actual or potential capacity to reason merit moral standing. Another commonly held position is that any beings that have the capacity to suffer

---

[1]One example might be to alter our central nervous system so as to reduce the likelihood that individuals would go into a rage when upset, or (perhaps even better) reduce the likelihood that individuals would get emotionally disturbed under adverse circumstances. Another example would be altering our central nervous system to make us become more caring and empathetic to other individuals. More mundane examples of enhancement might include improvements so that we retain facts with improved reliability, respond faster to dangerous or adverse situations or simply have improved physical strength.

[2]https://www.scu.edu/ethics/ethics-resources/ethical-decision-making/who-counts/.

ought to have some form of moral standing afforded to them that should at least make gratuitous cruelty against them immoral (Singer 1975; Cavalieri and Singer 1994; Singer 2011).

Peter Singer, in his seminal book *Animal Liberation* (Singer 1975), argues the case that animals do merit a degree of moral standing, with rights that extend far beyond merely preventing them from experiencing pain and suffering. For instance, he argues that humans are often guilty of "speciesism", the assigning of different moral standings to various biological entities on the basis of their species membership rather than on the basis of more appropriate properties such as their cognitive capacity.

3.3 Personhood

As noted earlier, the recognition of an entity as being a person is known as **personhood**. The philosophical literature offers many definitions as to what constitutes a "person", often drawing in large part on the work of philosophers such as Descartes, Locke and Hume, and usually focusing on criteria such as self-awareness, a sense of self that is maintained through time, or similar criteria.

One especially notable work in the study of personhood is Kim Atkins' *Self and Subjectivity* (Atkins 2005). This volume of classical philosophical writings with accompanying critical commentary examines how philosophers have historically conceived of personhood and the related notions such as self and subjectivity. Noting that humans, perhaps unique among beings, "have the capacity to reflect upon and evaluate their thoughts, feelings, and actions" and this ability for self-reflection "is the essence of philosophy" (Atkins 2005) the author offers the reader perspectives from philosophers such as Descartes, Locke, Hume, Kant, Nietzsche, Strawson, Frankfurt and Parfit and Simone de Beauvoir. Let us briefly examine some of these thinkers.

Descartes famously characterized the human subject in terms of the mutually exclusive realms of matter (*res extensa*) and thought (*res cogitas*), a position known as Cartesian dualism. This position is in contrast with (for example) physicalism, the ontological position that there is nothing over and above the physical and what we take to be mental events are simply "epiphenomena". (In physicalism, mental phenomena are seen as entirely dependent on physical events and consequently have no independent existence).

While the British philosopher John Locke is often regarded as the founder of empiricism, he also contributed greatly to the problems of personhood and personal identity, and his approach to the matter continues to be influential to this day. According to Locke, a living thing retains its personal identity by maintaining the organization of its component parts, that is, by maintaining a continued longitudinal existence. Locke defines a "person" as "a thinking intelligent being that has reason and reflection, and can consider itself as itself, the same thinking thing, in different times and places".[3] It is the Lockean approach to personhood that this work draws strongly upon, as is discussed in further sections.

In the "naturalist" epistemological tradition,[4] personhood is usually taken to designate any "agent" who is ordinarily in possession of continuous consciousness over time[5] and who also is capable of mental representations concerning the world, as well as able to consider options, develop tactics and act on plans.

Writing some decades after Locke, David Hume was a Scottish atheist and philosopher in the empiricist tradition who while best known for his attacks on religion, also contributed to the philosophy of personhood and personal identity. In his *Treatise of Human Nature* (1738) Hume attacked Descartes' notion of *res cogitas,* arguing as an empiricist that since we do not perceptually experience any metaphysical self we cannot have knowledge of it. Hume additionally argues that it is erroneous to posit an entity such as "the self" and that to do so is to mistake the connectedness of conscious experience for the existence of a soul.

Immanuel Kant, arguably the most important of the Enlightenment philosophers, continued the Lockean tradition of grounding philosophy in what we are able to know, especially in his famous book, *Critique of Pure Reason* (1781). Here Kant reconstructs Descartes's idea that knowledge is grounded in the human subject, arguing that while all knowledge begins with experience, it does not all arise out of experience. Thus Kant, like Hume, rejects Decartes notion of *res cogitas* but unlike Hume does not regard the self as a mere fiction, but instead as part of the composition of consciousness itself. That is, like most contemporary neuroscientists, Kant held that the composition of our conscious experience—and especially our sensory experiences—necessarily depends on the structure and function of the brain itself.

The philosopher Nietzsche rejected the Enlightenment ideal of reason guiding the progress of humankind, with its associated belief in self-sovereignty through reason. In fact, Nietzsche felt that an unwarranted emphasis on reason was responsible for a "herd-like" way of living and an associated decline in human creativity and individuality. In contrast with Decartes' *res cogitas* and *res extensa,* Nietzsche held that the notions of "mind" and "body" do not actually refer to distinct entities but are simply terms that help us refer to different aspects of our existence. His books *The Genealogy of Morals* and *The Birth of Tragedy* discuss his philosophical positions on these issues in great detail.

The Oxford philosopher Sir Peter Strawson offered a perspective on personal identity rather different from his predecessors. Strawson argued that metaphysics is concerned with the description of the basic structure of material existence, that is, the entities we take for granted in our concept of reality. He calls these entities

[3]John Locke, *An Essay Concerning Human Understanding*, II, ed. Raymond Wilburn, Chapter XXVII (London: Dent and Sons, [1690] 1948), p. 9. quoted in Atkins, 2005:20.

[4]Naturalistic epistemology is an approach to the theory of knowledge that emphasizes the application of methods, results, and theories from the empirical sciences. It contrasts with approaches that emphasize a priori conceptual analysis or insist on a theory of knowledge that is independent of the particular scientific details of how mind-brains work (From http://www.iep.utm.edu/nat-epis/).

[5]This phrasing allows for periods of unconsciousness, such as when general anesthesia is administered. Note that sleep is not taken to be an unconscious state.

"basic particulars" and includes "persons" among these (but not "bodies"), positing that persons are basic material entities with both physical and mental properties. Understandably, Strawson rejects the physicalist claim that mental states are mere predicates of the brain, arguing that psychophysical reductionism (incorrectly) describes all human activity in terms of the laws of physics. Strawson's nonreductionist approach nonetheless has the advantage of retaining the possibility of moral agency.

The Princeton University moral philosopher Harry Frankfurt has written extensively on the conceptual and moral aspects of personhood, especially as the notion pertains to moral responsibility. In his well-known essay, "Freedom of the Will and the Concept of a Person," Frankfurt argued that freedom of the will allows individuals to impose a moral structure on one's self through a system of held principles and values, and that those individuals who do not achieve self-control in this manner lack personhood. (He calls such individuals "wantons"). Frankfurt further argues that Strawson's concept of personhood is too broad to capture what is specific about persons, noting that some animals meet the definition of a being to whom both mental and physical predicates can be ascribed but nevertheless still do not have freedom of the will. Frankfurt emphasizes that critical self-reflection and self-determination based on a system of values are requirements for the moral fabric that distinguishes a person from a wanton. In this respect, only persons are moral agents, and so only persons are members of the moral community. Finally, it should be noted that the requirement for critical reflection as a necessary condition for personhood has met with criticism from philosophers who argue that such a requirement would exclude from personhood both infants as well as some neurologically damaged adults. In addition, this requirement is seen by some as doing an injustice to animal ethics.

The Cornell University philosopher Sydney Shoemaker, in an essay entitled "Personal Identity: A Materialist's Account" argues that persons exist as a "temporal whole", existing not at a single point in time, but rather extended over the entire period of their lives. To Shoemaker persons are divided into temporal parts ("time slices" or "person-stages"), which are the collection of properties attributable to a person at any time, so that a person is the totality of his or her existence across all that individual's person stages. Shoemaker offers this approach to avoid the continuity of consciousness problem[6] of the Lockean approach. In the Shoemaker approach, one need not establish that there is an identical consciousness stream at each stage of a person's life, but that there is instead continuity of memory connecting the person-stages in an individual's life.

Derek Parfit, at All Souls College, Oxford, is best known for his book, *Reasons and Persons*. Like Shoemaker, Parfit considers personal identity to consist in memory-connectedness. However, unlike Shoemaker and many other philosophers, Parfit offers us a reductionist position that does away entirely with the concept of "person".

---

[6]Simply stated, the problem is this: if, as Locke holds, personhood depends on consciousness, what happens to personhood when one's body relinquishes consciousness, such as with general anesthesia? Additionally, since personal identity is based on consciousness, and that only oneself can be aware of his consciousness, exterior human judges may never know if they really are judging—and punishing—the same person, or simply the same body.

His position is based on a fascinating thought experiment involving teleportation (as in the *Star Trek* series); the interested reader is directed to Parfits' book and Atkins' commentary (Atkins 2005) for further explanation.

Finally, it is of interest to explore one of the feminist perspectives on personhood, in this case that of Simone de Beauvoir. In *The Second Sex* (1949), Beauvoir notes that a woman is defined only by her relation to men, whereas a man is defined in terms of himself standing alone. Similarly, what men carry out is routinely assigned legitimacy, but the actions of woman require justification by men. This arrangement of "physiological destiny", Beauvoir argues, leaves men free to pursue authentic, self-affirming activities while trapping women in a setting of domestic imprisonment. Beauvoir's concerns, however, should not be interpreted to means that women do not merit full-fledged personhood, only that women often find themselves trapped in situations where they are not able to achieve their full potential.

As can be seen from the foregoing, a variety of different perspectives on personal identity have been advanced by philosophers. Let us see how such notions might be extended to nonhuman organisms.

Hughes (2004) has extended the work of Locke and others in the Lockean tradition to consider a spectrum of rights in organisms varying in the extent to which they exhibit consciousness and other morally important properties (Table 3.1). Four degrees of personhood are described in his scheme: individuals with mature personhood (e.g., adult humans), individuals with personhood (children, some apes), beings with sentience (e.g., fish), and nonsentient entities (e.g., embryos and plants).

Some philosophers argue that the classic view of personhood identified above suffers from the problem that it defines personhood in terms of an ability to perform certain tasks. This approach makes it difficult to distinguish between persons as we understand them intuitively and advanced machines capable of complex adaptive behavior. Consequently, some philosophers such as Charles Taylor propose that personhood be based on a view emphasizing **personal significance** rather than on function (Taylor 1985):

> What is crucial about agents is that things matter to them. We thus cannot simply identify agents by a performance criterion, nor assimilate animals to machines… [likewise] there are matters of significance for human beings which are peculiarly human, and have no analogue with animals.

Other philosophers have similarly argued against a view of personhood based on function. For example, philosopher Harry G. Frankfurt argues that the concept of a person is closely connected to human volition, and describes the makeup of human volition in terms of first- and second-order desires (Frankfurt 1971):

> Besides wanting and choosing and being moved to do this or that, [humans] may also want to have (or not to have) certain desires and motives. They are capable of wanting to be different, in their preferences and purposes, from what they are. Many animals appear to have the capacity for what I shall call "first-order desires" or "desires of the first order," which are simply desires to do or not to do one thing or another. No animal other than man, however, appears to have the capacity for reflective self-evaluation that is manifested in the formation of second-order desires.

Table 3.1 Modification (elaboration) of Hughes's classification of the future continuum of consciousness and rights in organisms (Hughes 2004)

Type of Organism	Consciousness	Rights
• Conventional adult humans • Cognitively enhanced humans ("Post-humans") • Fully sentient androids and cyborgs with full adult human intellectual capabilities • Adult human clones	Full personhood	• Right to vote • Right to make contracts • Right to humane treatment
• Young children • Adult great apes • Human adults with impaired cognition • Selected mammals (e.g. Dolphins)	Limited personhood	• Full right to life • Right to humane treatment
• Newborn infants • Adults with severe cognitive impairment • Fish and birds • Human fetuses at the stage of advanced neuronal development	Sentience (*the capacity to feel, perceive or experience subjectively but not necessarily able to reason*)	• Limited right to life • Right not to suffer unnecessarily
• Permanently unconscious humans • Embryos and early fetuses • Brain dead humans • Anencephalic newborns • Anencephalic or headless human clones • Plants	Non-sentient (*unable to feel, expeience or suffer*)	• No rights: property of family or owners (in case of plants)

Philosopher Thomas I. White, in developing an argument that dolphins are "non-human persons", argues that the criteria for personhood should be as follows (White 2007): the candidate being (1) is alive, (2) is aware, (3) feels positive and negative sensations, (4) has emotions, (5) has a sense of self, (6) controls its own behavior, (7) recognizes other persons and treats them appropriately, and (8) has a variety of sophisticated cognitive abilities.

Beauchamp, on the other hand, cautions against approaches such as those listed above. Emphasizing that moral personhood is quite distinct from metaphysical personhood (as only the former demands moral agency and moral motivation), he writes (Beauchamp 1999):

> The belief persists in philosophy, religion, science, and popular culture that some special cognitive property of persons like self-consciousness confers a unique moral standing. However, no set of cognitive properties confers moral standing, and metaphysical personhood is not sufficient for either moral personhood or moral standing. Cognitive theories all fail to capture the depth of commitments embedded in using the language of "person".

Beauchamp's concerns notwithstanding, the traditional Lockean approach (or variations thereof) to discussing personhood remains an important, and influential philosophical tool accepted by a great many moral philosophers including individuals such as Hughes and others discussed above.

3.4 Nonhuman Personhood

To many philosophers, drawing on the above Lockean criteria for personhood, persons may not necessarily be humans (some animals may merit personhood) while some human entities such as embryos, fetuses or severely brain damaged adult patients may not qualify for personhood. As examples, under such a scheme, some of the great apes may qualify for personhood while some people afflicted with severe dementia from Alzheimer's disease or traumatic brain injury may not.

Closely related to the discourse on what constitutes personhood are issues of moral standing, individual rights and ethical responsibility. For instance, it is commonly held that all individuals meeting personhood criteria are deserving of at least some individual rights and that only persons (and not, for instance, lower animals) are expected to be ethically responsible for their actions. Another ongoing philosophical debate concerns whether people (who may or may not be persons) deserve greater rights and higher moral standing than non-people; the rights of nonhuman animals are an example (Howard 2005; Sagoff 2007).

These issues are more than theoretical. For instance, it is now generally accepted that many Great Apes can learn to communicate with humans using signs and symbols (such as using lexigrams or American Sign Language), although the exact extent of their capabilities still remains a matter of ongoing debate among anthropologists and primatologists. As a consequence of new considerations such as the above, the Great Ape Project has been launched. With an online presence at http://www.greatapeproject.org, the Great Ape Project is an international organization of animal advocates who are actively promoting the position that the United Nations should pass a "Declaration of the Rights of Great Apes" that would confer basic legal rights to chimpanzees, bonobos, gorillas, and orangutans. These proposed rights would include the right to life, protection of individual liberty, release from captivity, and protection from torture.

Two recent legal cases exemplify developments in this arena. In an Argentinian court decision rendered on December 18, 2014, a great ape living in Buenos Aires was recognized as having the same basic rights as her human captors. The female orangutan, living in the Buenos Aires Zoo for the last two decades, now has the rights of life, liberty and freedom from harm.[7]

Similarly, in the USA, in 2013, the Nonhuman Rights Project launched three legal claims on behalf of four chimpanzees in New York State.[8]

In this and earlier sections of this chapter we examined some of the issues pertaining to personal identity, moral standing and personhood. In the following sections of this chapter we use this background to additionally explore how various forms of human biological manipulation (e.g., eugenics, and genetic engineering) also impacts on these same concerns.

3.5 Classical Eugenics

Eugenics is a social philosophy that takes the position that the human gene pool can be improved through interventions targeted at the "genetically unfit" via techniques such as sterilization, selective breeding, and (sometimes) even euthanasia (Kevles 1985; Black 2003). Eugenics has also been defined as the self-direction of human evolution.

A line of argument frequently raised against the visions and goals of those who would improve the design of the human body by techniques such as genetic engineering is concern for abuses that have occurred in the past in the name of eugenics. One need only look to the state-sponsored murdering of millions of Jews by the Nazis in the name of racial purity to be wary of any mention of eugenics. In more recent times, "ethnic cleansing" in the Balkans or in Myanmar has led to similar concerns.

Finally, other critics point to sterilization programs in the USA and elsewhere that were aimed at eliminating genetic defects from the gene pool. For instance, Margaret Sanger, an American feminist well-known as an early advocate of birth control, argued for: "A stern and rigid policy of sterilization and segregation to that grade of population whose progeny is already tainted or whose inheritance is such that objectionable traits may be transmitted to offspring" (Sanger 1932).

3.6 Liberal Eugenics

Given the obvious ethical lapses of classical eugenics, some thinkers have considered how the aims of classical eugenics might be met without moral peril. This is the aim of liberal eugenics. The central notion in this instance is that society should non-coercively employ modern biological technologies to enhance the human organism and its genetic composition. As noted above, the term is used in distinction from

---

[7] http://www.scientificamerican.com/article/argentina-grants-an-orangutan-human-like-rights/.

[8] http://www.nonhumanrightsproject.org/2015/04/20/judge-recognizes-two-chimpanzees-as-legal-persons-grants-them-writ-of-habeas-corpus/.

the classical "authoritarian" eugenics programs of the early and mid-1900s, which used unethical coercive methods to decrease the frequency of "undesirable" hereditary traits via programs of racial hygiene, sterilization, extermination, and ethnic cleansing. In contrast to these "negative" means of improving the human condition,[9] liberal eugenics focuses on the "positive", with its actions and goals carried out by individuals ethically exercising "procreative liberty" with a view to achieving "procreative beneficence". For instance, Nicholas Agar's 2004 book *Liberal Eugenics: In Defence of Human Enhancement* (Agar 2004) he argues the case that parents should be permitted to enhance their children's genetic composition via available biological methods and argues against often-raised concerns that such actions would "divide society into genetic haves and have-nots" or would "lead to a single view of human excellence".

Unlike some transhumanists, Agar does not argue that we have a moral duty to employ enhancement technologies to improve the human condition. Nor does he offer arguments based on any particular philosophical school, such as the teachings of Immanuel Kant or any utilitarian philosophers. Rather, he simply argues that to maintain consistency with the prevailing liberal democratic values of freedom of choice and self-determination, we must agree to the deployment of human enhancement technologies by parents wishing to do so.[10]

Similarly, in 2006 Johann Hari, a writer for the British newspaper *The Independent* offered the following argument in favor of liberal eugenics[11]:

> The criticism that deserves more careful consideration comes from disability rights activists like Adrienne Asch. They argue that this attempt to eradicate disability is an assault on disabled people. By trying to eradicate disabilities, we are saying disabled people are worth less – "errors in the gene pool" - and clearing the way for them to be treated even more badly.
>
> But is this true? By making sure that no more mothers take thalidomide during pregnancy, are we implicitly saying that thalidomide people have worthless lives and should be killed? Of course not. We are simply saying that a person is more likely to be able to live the kind of life they want to with fully formed arms and legs. By ensuring that the number of able-bodied babies are maximized, we are simply acknowledging that - however harsh it might seem to say it - lacking an ability to hear or see or walk is not simply a difference. It is a disability nobody would voluntarily choose, and that you are better off without. Nor does the evidence suggest greater screening and treatment will lead to the remaining disabled people being treated worse. Since amniocentesis was introduced, people with Down syndrome are, if anything, treated better.

In essence, Agar, Hari and similarly minded individuals argue that acting proactively to best ensure that all babies are the best that they can be, for example free of medical problems that would be expected to limit their life choices, is an option

[9]For a more detailed discussion of what I mean by "human condition" see Chap. 4.

[10]That being said, the state may choose to limit what a parent may do on behalf of his or her child on the basis of safety, because of a concern about child exploitation, or for other reasons. For example, in the interests of safety governments may mandate that a child be vaccinated against a newly emerging virus or may limit the extent to which children are allowed to work outside the home to help prevent child abuse.

[11]http://www.independent.co.uk/voices/commentators/johann-hari/johann-hari-why-i-support-liberal-eugenics-8689688.html.

that should be available to future parents, at least in principle.[12] This is not to suggest that the humanity and value of disabled individuals are less in comparison to other individuals, just that freedom of choice, independent of any particular ethical or religious framework, should be afforded to parents who want to provide the best opportunities for their babies.

3.7 Introducing Transhumanism

Transhumanism is a relatively recent intellectual and cultural movement that promotes an interdisciplinary approach to understanding and evaluating potential means for enhancing the human organism, the human race and the human condition, as well as removing some of its limitations. The potential goals of transhumanism are numerous and diverse, but as a rule, transhumanism seeks to produce "posthumans" who will variously live longer, be smarter, be stronger, be more peace-loving, or otherwise constitute an improvement over regular humanity in some nontrivial manner. In this sense, transhumanism is rather similar to the field of liberal eugenics discussed earlier. A more complete understanding of the goals of transhumanism can be achieved by studying the **Transhumanist Declaration**, displayed as Table 3.2, as well as by visiting their website at http://humanityplus.org.

Transhumanism aims to achieve its goals primarily but not exclusively via technological innovation in the fields of genetic engineering, neuropharmacology (as well as other branches of pharmacology), computer technology, artificial intelligence, molecular nanotechnology and a number of other scientific domains.

Some of the less mainstream transhumanist themes include efforts to develop conscious, self-aware computers, efforts to reanimate individuals deemed to be dead by ordinary clinical criteria, and altering humans with a view to facilitating space colonization.

For example, a few transhumanists are involved in the cryonics movement, which involves attempting to freeze (cryopreserve) humans so that they can be thawed in the future, usually with a view to obtaining advanced medical treatment not yet available. So confident are some individuals that continuing medical advances will eventually lead to cures for most diseases that they want to have their bodies frozen just before dying of a terminal illness. (For legal reasons, however, the clinically less desirable option of being frozen a brief period after dying is a far more practical option. However, it should be noted that the application of cryonics to patients who are already clinically dead is a particularly controversial aspect of cryonics, as many authorities compare the use of cryonics in this setting to the selling of false hope in the extreme).

A basic undercurrent in most transhumanist writings is a vision of wisely using human ingenuity to develop technology to improve the human condition, a vision

---

[12] Real-world limitations such as cost and safety considerations may, however, limit the extent that such interventions are practical.

Table 3.2 The Transhumanist Declaration, 2009 revision. World Transhumanist Association

	The Transhumanist Declaration
1.	Humanity stands to be profoundly affected by science and technology in the future. We envision the possibility of broadening human potential by overcoming aging, cognitive shortcomings, involuntary suffering, and our confinement to planet Earth
2.	We believe that humanity's potential is still mostly unrealized. There are possible scenarios that lead to wonderful and exceedingly worthwhile enhanced human conditions
3.	We recognize that humanity faces serious risks, especially from the misuse of new technologies. There are possible realistic scenarios that lead to the loss of most, or even all, of what we hold valuable. Some of these scenarios are drastic, others are subtle. Although all progress is change, not all change is progress
4.	Research effort needs to be invested into understanding these prospects. We need to carefully deliberate how best to reduce risks and expedite beneficial applications. We also need forums where people can constructively discuss what should be done, and a social order where responsible decisions can be implemented
5.	Reduction of existential risks, and development of means for the preservation of life and health, the alleviation of grave suffering, and the improvement of human foresight and wisdom should be pursued as urgent priorities, and heavily funded
6.	Policy making ought to be guided by responsible and inclusive moral vision, taking seriously both opportunities and risks, respecting autonomy and individual rights, and showing solidarity with and concern for the interests and dignity of all people around the globe. We must also consider our moral responsibilities towards generations that will exist in the future
7.	We advocate the well-being of all sentience, including humans, non-human animals, and any future artificial intellects, modified life forms, or other intelligences to which technological and scientific advance may give rise
8.	We favour allowing individuals wide personal choice over how they enable their lives. This includes use of techniques that may be developed to assist memory, concentration, and mental energy; life extension therapies; reproductive choice technologies; cryonics procedures; and many other possible human modification and enhancement technologies

Source http://humanityplus.org/learn/transhumanist-declaration/

free of concerns about whether "playing God" or deeply altering our human character is ethically problematic. Not surprisingly, however, it is exactly such concerns that critics of transhumanism raise. These issues are discussed in later sections of this work. But first, let us briefly look at three important contemporary philosophers who are especially sympathetic to transhumanist principles.

3.8 Nick Bostrom

Nick Bostrom is director of The Future of Humanity Institute at Oxford University as well as the author of a vast number of publications on a diverse variety of philosophical topics (e.g., existential risk, implications of consequentialism, the simulation argument, transhumanist issues). However, it is his interest in transhumanism (he co-founded the World Transhumanist Association) that primarily interests us here. Bostrom has made significant contributions to such transhumanism-related topics such as human enhancement, artificial intelligence, superintelligence, mind uploading, nanotechnology, and cryonics. His anthology, *Human Enhancement* (Savulescu and Bostrom 2008) provides a comprehensive look at the question of human enhancement by artificial means, with commentaries by authorities from both transhumanist and bioconservative ("anti-meliorist") camps.

3.9 Julian Savulescu

Julian Savulescu is Uehiro Professor of Practical Ethics at the University of Oxford as well as Editor-in-Chief of the *Journal of Medical Ethics*. His approach to bioethical issues is principally utilitarian. Savulescu argues that stem cell research requiring embryo destruction is justifiable even if the embryo is seen as a person (Savulescu 2002), that parents have a responsibility to select the best children they are able (procreative beneficence) (Savulescu 2001b), and that the destruction of embryos and fetuses as a source of material for transplantation to adults can be morally acceptable (Savulescu 1999a). Savulescu also argues that genetic testing for genes for criminality is a moral imperative (Savulescu 2001a) and that—based on respect for procreative autonomy—a strong argument exists to allow parents to select the sex of their children (Savulescu 1999b; Savulescu and Dahl 2000).

3.10 John Harris

Bioethicist John Harris, Sir David Alliance Professor of Bioethics at the University of Manchester, is perhaps best known for his book *Enhancing Evolution* (Harris 2007), in which he makes a well-articulated case in support of genetic enhancement, designer babies, stem-cell research, and even human cloning. Previous books by Harris include *On Cloning* (Harris 2004) and *Clones, Genes, and Immortality* (Harris 1998), both of which, like *Enhancing Evolution* take a libertarian-utilitarian-consequentialist position advocating that humanity should act to improve the human organism (and in so doing, the human condition) through science, technology and imagination. Harris argues that science is a moral good that should be harnessed to improve humanity, and not repressed as many detractors have attempted to do[13] (Harris 2005a). He has also argued in favor of sex selection (Harris 2005b, c), argued for the acceptability of using drugs like methylphenidate to enhance performance (Harris 2009), and suggested that drugs might also be helpful in the moral enhancement of humanity (Harris 2011; Chan and Harris 2011). Harris has even suggested that it could be ethical to sell organs in the marketplace (Erin and Harris 2003).

As mentioned above, Harris justifies his arguments by drawing on libertarian-utilitarian-consequentialist lines of ethical reasoning, stating that scientific developments that would be expected to improve humankind should be encouraged, even if they might initially appear to be repugnant or repulsive. Harris argues that we have a "positive duty" to enhance humankind in that enhancements, by definition, are good. In particular, Harris supports this claim by stating that—drawing on the principle of beneficence—we have a moral obligation to prevent forms of serious harm such as those that genetic diseases so often cause. Thus to Harris, the cloning of human organisms, while widely condemned or outlawed, has the potential for bringing about good, for example as a research technique to understand and treat human hereditary diseases. Consequently, Harris argues, the cloning of human organisms should not be universally forbidden.

Harris argues that gut feelings are poor guides to moral decision making, that personal prejudices are a poor way to address life's thorny questions. Harris notes that just as "test-tube babies" are now widely accepted despite an initial outcry, we can expect that various forms of human enhancement will eventually gain acceptance as well once their benefits to humankind become fully appreciated.

Critics of Harris generally offer arguments along several lines. Some, like Kass (see Sect. 7.4), suggest that repugnance *can* be a guide to moral decision making. Some others argue that the utilitarian-consequentialist ethical model that Harris draws on is too fundamentally flawed to serve as a moral compass, especially in matters involving the "sanctity" of human life, where a duty to respect and preserve human life is fundamental (see, for instance, Sect. 2.2). For additional critical commentary the interested reader is directed to detailed discussions by Brassington (2007), Shapshay and Pimple (2007) and Stjernschantz Forsberg et al. (2014).

---

[13]For an American perspective on how political forces sometimes act against scientific interests, the reader is invited to read Mooney's *The Republican War on Science* (2006).

3.11 Charles Tandy

Another important contemporary Transhumanist thinker is Charles Tandy, whose research interests include the possibility of immortality via interventions such as genetic engineering, nanotechnology and cryonics. Tandy places a particular emphasis on Robert C. W. Ettinger's monograph, *The Prospect of Immortality* (Ettinger 2005a) and his book, *Man Into Superman* (Ettinger 2005b) as two of the most important early works in this realm. Tandy also provides valuable insights into Ettinger's philosophy in his 2002 book *The Philosophy of Robert Ettinger* (Tandy 2002) as well as in a number of his other works readily accessible via his website (DoctorTandy.com).

3.12 Enhancing Human Performance for Spaceflight

As an interesting example of one of many possible transhumanist themes, consider how one might enhance human performance for spaceflight, for instance using pharmacologic enhancement. (Gibson 2006) has considered the ethical issues in enhancing human performance in such a setting. For instance, for decades, military aircrews have employed pharmacologic assistance using drugs such as amphetamines to aid in completing their missions. While this is sometimes said to be done only under medical supervision and only under wartime circumstances, it is a well-documented practice (Kenagy et al. 2004). Consequently, one could also imagine that similar circumstances might arise in space travel, and as a result, considerable study has gone into exploring just how drugs might be wisely deployed in such a setting. (We will look at the issue of pharmacologic enhancement in detail in the next chapter).

Similarly, just as scientists and military personnel involved in Antarctica missions are sometimes required to have a prophylactic appendectomy before beginning a mission where they would be snowbound for extended periods (Schmidt 1965), such a policy might also be sensible prior to long duration spaceflights, such as a mission to Mars. In the same way, corneal remodeling surgery, such as the Lasik laser eye procedure, may be valuable in such a setting.

At issue, however, in such cases, is how extensive such surgical procedures or other interventions might be before becoming unethical. As an extreme case in point, Gibson provides the example of a bilateral leg amputation that might be carried out in order to "enhance tolerance to gravitational forces". Similarly, the question arises concerning the point at which the surgical implantation of devices (or similar interventions) to enhance human performance would become unethical.

In a comparable manner, while the genetic modification of an existing individual or an unborn individual to cure an ailment might be considered ethically acceptable for present or future astronauts, Gibson points out that conducting genetic modifications to alter an individual's performance characteristics and personality traits is less likely to be acceptable to most people.

While it could be argued that all that is ethically germane in such situations is that all participants provide informed consent, this may not be enough, as in the very controversial example of individuals from developing countries who would consent to sell a kidney to help feed their family (Epstein 2007; Friedman and Friedman 2006). Almost all bioethicists maintain that it is unethical for one to sell a kidney even when complete informed consent is present.[14]

3.13 Animal/Human Hybrids and Other Strange Creatures

Some scientists are now putting human genes into animals for research and clinical purposes.[15] Others are thinking about why one might want to put animal genes into humans. Others are working on the development of part-human, part-animal cells and embryos for research purposes. Such considerations raise issues as to what it means to be a human. If we put human genes into animals, does that make the animals more human? Is it morally wrong to produce animal/human hybrids? In the following sections we begin an exploration of these issues.

3.14 Putting Human Genes into Animals

For some time, transplant surgeons have been interested in using animal organs in human hosts (xenotransplantation) as an alternative to transplanting human organs (Cox and Zhong 2005; Rowiński 2007). The primary motivation for this is a shortage of human organs available for transplantation. Unfortunately, attempts at using animal organs for transplantation in humans have met with universal failure in that organ rejection is almost inevitable. This rejection occurs even when employing aggressive anti-rejection therapy and occurs as a direct result of immunological differences between animals and humans.

The basis for this is as follows. All cells in humans and animals have on their surfaces molecular markers that can be recognized by the immune system. If the immune system sees markers that indicate that the cells being scrutinized are "foreign", a complex series of biological actions are initiated whereby the foreign material is attacked. When the material being attacked is a transplanted organ, organ rejection occurs.

Because the immunological markers located on cell surfaces are coded by a cell's DNA, consideration has been given to producing genetically modified animals whose

[14] A notable exception is John Harris (Erin and Harris 2003).

[15] One of the most important recent developments in is this field is a radical new gene-splicing technique called "clustered regularly interspaced short palindromic repeats" (CRISPR). CRISPR is a new tool to edit DNA with unprecedented precision. For more information, a highly readable account is available at http://gizmodo.com/everything-you-need-to-know-about-crispr-the-new-tool-1702114381.

cell surface markers will not trigger an immunological response. One proposed scheme is to place appropriate human DNA in an animal's genome such that the cell's surface markers indicate that the cell is human rather than, say, porcine in origin, fooling the new system in a "fox in sheep's clothing" manner.

Another situation where scientists may wish to place human DNA is in an animal to produce pharmaceutical agents. For instance, one might wish to produce cows whose milk is rich in human insulin, human growth hormone or somatostatin. This is similar to earlier research programs that had bacteria produce human insulin and other agents.

3.15 Language Genes

In 2001, scientists studying a family in which many members were unable to speak identified a gene central to spoken language (Lai et al. 2001). The gene, known as FOXP2, is not unique to humans; however, humans have a unique *form* of FOXP2 that is essential to language production (Vernes et al. 2006). For instance, the human variant of the gene differs at two critical points from the chimpanzee FOXP2 gene, implying that these changes may be responsible for the fact that humans can speak while chimpanzees cannot. Such a situation makes one wonder whether primates provided with the human form of the FOXP2 gene via genetic engineering techniques might be made to exhibit a degree of language production superior to that of ordinary ("unmodified") animals. Of interest, in a recent study of mice engineered to express humanized Foxp2, the mice learned to run a maze much more quickly than normal mice and were also better at turning mindful actions into behavioral routines.[16]

3.16 Human/Animal Chimeras

A chimera, named after a monster in Greek mythology that had a lion's head, a goat's body, and a serpent's tail is a mixture of two or more species in one organism. Scientists have now started to blur the once clear line between what is human and what is animal by producing chimeras that are part human and part animal. This process is distinct from the gene transfer methods alluded to above in that it involves fusing two entirely different cell types.

For instance, scientists at the Shanghai Second Medical University have fused human cells with rabbit eggs as part of a means to produce stem cells while at Stanford University's Institute of Cancer/Stem Cell Biology and Medicine in California, using somewhat similar techniques, Irv Weissman has created mice with brains that are about one percent human and plans to eventually produce mice that have 100%

[16]http://newsoffice.mit.edu/2014/language-gene-0915.

human brains.[17] The process of producing these chimeras involves injecting human neurons into the brains of embryonic mice. With a view to understanding the process of neuronal development, before being born the mice are killed and dissected to determine if the neuronal architecture appropriate to a human brain has formed.

As implied above, it is usually not the intent that such embryonic organisms be brought to term but rather that they be destroyed after a particular number of cell divisions. Arguments in favor of developing such hybrids center on their utility in biological research. Arguments against often present a "sanctity" or "dignity" argument about human life and its pricelessness but may also focus on safety issues that apply when generating new life forms.

As a rule, the more genetically human-like the research animals are, the better scientists will be able to study the progression of human diseases in animal models, as well as test new drugs for efficacy and toxicity. Thus some investigators are considering the development of human-chimpanzee chimeras. Given that chimpanzees are already very closely related to us, sharing about 98% of the human genome,[18] adding the right additional genetic material by such means as fusing a human and a chimpanzee embryo might be expected to produce a perfect test vehicle for research into human disease, as well as provide an excellent source of organs and tissue for transplantation purposes. But naturally, the development of such a "humanzee" would raise difficult ethical questions. Should such an organism enjoy some rudimentary form of human rights? Might it be forced into doing menial labor or even used to perform dangerous jobs, like cleaning up nuclear spills?

A related concern would be that of creating an organism with human consciousness trapped in an animal body, perhaps directly or indirectly as a result of mating two such especially endowed animals. While admittedly far-fetched to some individuals, from a scientific point of view the possibility of this happening is less than remote and requires critical consideration.

As a result of such issues, some scientists and scientific organizations such as the National Academy of Sciences have developed stem cell research guidelines that prohibit the breeding of animals that carry human stem cells. As a further measure, some ethicists and scientific organizations would forbid the transplantation of human stem cells into higher primates like chimpanzees to even further reduce any chance of producing a humanized animal with human intelligence. Because animals that have been "humanized" by stem-cell transplant techniques may offer excellent opportunities for testing new drugs and therapies, such research initiatives will continue to be developed. Presumably, however, the more human content is present in such animals,

---

[17]Mott M. Animal-Human Hybrids Spark Controversy. National Geographic News. January 25, 2005. Accessed from http://news.nationalgeographic.com/news/2005/01/0125_050125_chimeras.html. See also the following account from 2013 by Jonathan Moreno entitled "Researchers Put Human Brain Cells in Mice. Why Aren't Conservatives Freaking Out?", available online at http://www.slate.com/blogs/future_tense/2013/03/15/why_aren_t_conservatives_upset_about_researchers_putting_human_brain_cells.html.

[18]See, for instance, http://www.ornl.gov/sci/techresources/Human_Genome/faq/compgen.shtml.

the more likely they are to be a useful model to study human diseases. There may thus be a temptation to continue to add human material until the end result is more human than animal, which of course brings all the problems of moral standing in its wake.

3.17 Inserting New Genes in Existing Humans

If inserting human genes into animals is morally intriguing, the idea of inserting new genes (perhaps animal genes) into humans is surely equally intriguing. Gene therapy involves adding one or more genes to a person's genome for some therapeutic purpose, such as ensuring that an important enzyme is produced (Mammen et al. 2007).[19] Viruses are sometimes used to transport the extra DNA into cells, a process that is still experimental and, as described in Sect. 7.13, occasionally hazardous.

As gene therapy improves its safety profile and becomes better understood it is envisioned that treatments will become available for a number of genetic afflictions. Perhaps extra copies of some special genes might be added with a view to providing extra strength, extra brain power, or whatever is associated with the special added gene.

Consider the following (so far) hypothetical example. Humans possess somewhere between 20,000 and 25,000 different genes. But if some individuals were to possess multiple copies of a particularly desirable gene in their genome, the expected result would be that more protein coded by the gene would get produced. In the case of the protein coded for by the AQP7 gene, which functions as an important element in transporting materials across the cell membrane, the protein allows water and glycerol to enter the cell, where they are used to produce energy. It is thus speculated that having multiple copies of the AQP7 gene might be valuable in endurance athletics such as marathon running, when the body needs to mobilize energy molecules from fat stores.

3.18 Human/Computer Hybrids and Neuroprostheses

For some time, futurists and science fiction writers have described how direct computer-brain interfacing might be used for human enhancement. In fact, in recent years significant advances in this general area have been made, with neuroprosthetic developments such as cochlear implants for deaf individuals (Papsin and Gordon

[19]An example would be a treatment for Cystic Fibrosis. This disease is caused by a defective gene encoding a protein known as *cystic fibrosis transmembrane conductance regulator* (CFTR), and is characterized by chronic lung infections and progressive lung damage that ultimately results in early death (Lee and Southern 2013).

2007), spinal cord stimulators for the treatment of intractable pain (Kuchta et al. 2007), and deep brain stimulators for managing Parkinson's disease (Starr et al. 1998).

In addition, research into retinal prostheses (Mokwa 2007), deep brain stimulation for the treatment of depression (Larson 2008), and even for drug addiction treatment via electrical stimulation of the nucleus accumbens(Van Kuyck et al. 2007) is taking place. In particular, research into "neuro-chips" where living brain cells and silicon circuits are coupled together shows special promise. It is hoped that this technology will lead to the development of improved neural prostheses, allow scientists to evaluate the in vitro effects of new drugs on neurons, and perhaps even lead to hybrid computers that will operate using living neurons in conjunction with ordinary silicon integrated circuits. Regardless, it is entirely reasonable to expect that these technologies might be employed to enhance human capabilities rather than just treat neurological disabilities and other clinical conditions.

The possibility that such technologies might also be employed for mind-control purposes must also be considered (Delgado 1969). However, because these technologies have not been widely advocated as a means to enhance human capabilities, they have stayed well under the "social radar". Nevertheless, the ethical issues involved with these technologies are not substantially different than those associated with the enhancement of human capabilities by other means, such as by way of pharmacological efforts, by genetic manipulation, or by surgical intervention. In addition, just because individuals with electronic neural prostheses are "partly electronic" should not make them any less human; they would still pass the usual criteria for humanness, such as DNA makeup, the ability to reason, and so on. And because they meet the usual criteria for humanness, they necessarily merit the moral standing given to human persons. This includes non-derogable rights (the right to life, the right to be free from slavery, the right to be free from torture and the right to be free from retroactive application of penal laws) as well as other rights, such as freedom of speech and freedom of movement. [20]

Although it should be obvious that human persons enhanced by pharmacological means, by genetic manipulation, by surgical intervention or by other means still merit the moral standing deserving of all human persons, the issue becomes less obvious when we consider the more radical idea of "artificial beings" who may have a physical embodiment completely different from human beings, such as being made mostly or even entirely of silicon circuitry. This is considered next.

[20]The fact that many nation-states do not honor these rights in no way diminishes the fact that human persons are morally deserving of a number of rights and freedoms; this fact merely points to the moral corruption of certain nation-states. An additional issue concerns the debate between scholars who advocate moral relativism and those who argue for moral universalism. The former argue that human rights are culturally contingent social constructs while the latter hold that human rights have always existed, and apply to all people regardless of ethnicity, race, culture, education, sex, or religion.

3.19 Robots, Androids and Artificial Persons

As noted in the introduction to this book, for many decades, science fiction writers have written stories about sentient computers, about intelligent robots serving humans as companions, about cyborgs made from a blend of human and machine parts, and about fully organic humanoids who were not at all human in any traditional sense. As today's scientists draw on developments in computer technology, robotics, artificial intelligence, molecular neuroscience, nanotechnology, and other emerging fields it is reasonable to expect that some of these notions will gradually move from science fiction towards scientific reality. There are already quite a few examples that could be cited to support such a trend.

One ongoing scientific effort on the fringes of science and engineering is to attempt to create sentient artificial beings embedded with consciousness. One particularly interesting philosophical issue in this context is whether a sufficiently complex machine might ever become self-aware. Most people agree that the "seat" of consciousness lies in the brain. And virtually all neuroscientists attempt to explain the workings of the brain as if the brain were a machine, albeit very possibly the most complex machine in existence.[21] But if the brain is just a conscious machine governed by physical laws, could it not be possible, at least in principle, to construct a computer that is conscious just like a biological brain. Wooldridge (1968) expresses this point as follows:

> … all intelligence whether of computer or brain, is a natural consequence of the powerful symbol-manipulating capabilities of complex switching networks and therefore the ordinary laws of the physical scientist are adequate to account for all aspects of what we consider to be intelligent behavior.

Wooldridge (1968) also takes the position that consciousness is determined by physics:

> The content of consciousness, as well as its presence or absence, is determined in detail by the physical structure and electrochemical state of the material of the brain.

He later elaborates on this issue (p. 159):

> The nonscientist may feel that there is something unnatural about the suggestion that consciousness may be suitable for absorption into the subject matter of physics. It may even seem to him that even if the lawful relations between the content of consciousness and the physical state of matter are eventually established, this will not really explain consciousness; we still don't know what it is or where it comes from. And this is true - nothing yet proposed provides any hope of explaining consciousness in any fundamental way. Further, it is safe to predict that no such explanation will ever be discovered. But precisely the same statement could be made about mass, electricity, or gravity.

---

[21]Note that some thinkers suppose the computer model to be a poor fit to understand human consciousness. In his 1989 book T*he Emperor's New Mind*, Roger Penrose speculated that quantum mechanical processes inside individual neurons gave humans special cognitive advantages over machines. Hubert Dreyfus has argued that human intelligence and expertise depends primarily on unconscious instincts rather than conscious symbolic manipulation, and that these unconscious skills cannot be captured as formal (algorithmic) rules employed by contemporary computers (Dreyfus 1979, 1986, 1992).

Arguably, the question as to whether machines can ever be conscious may be related to the question whether consciousness in the brain is active or passive. A ***passive consciousness*** is one which, while possibly arising as a natural consequence of a particular structural organization, nevertheless does not causally influence any material entities (such as neurons). Rather, it is a passive bystander, a result, rather than a cause. By contrast, an ***active consciousness*** is one which can actively "will" events, one which can be an initiating cause in a sequence of neuronal events. The active consciousness position views consciousness as an actual causal agent, one which allows for free will.

Although most neuroscientists do not write about consciousness issues explicitly, they most certainly marvel over the mystery of consciousness. However, most adopt the position that consciousness is passive in nature, implicitly allowing for the scientific investigation of behavior and intelligence without any consideration of consciousness. Simply put, while physical processes in the brain are held to exert considerable influence over consciousness, the passive consciousness position holds that consciousness does not exercise influence over physical activities within the brain. This is related to the position of "Strong AI", discussed next. Free will is not held to exist in such a setting.

Regardless of this issue, since the physical basis of consciousness is not yet understood except possibly in the broadest of strokes, many scientists regard the goal of producing an artificial sentient being as unachievable in the near future, if at all. Indeed, some philosophers such as McGinn, in his 1999 book *The Mysterious Flame: Conscious Minds in a Material World*, regard the problem of consciousness to be too intractable for the best of ordinary human brains to solve, just as a cat surely does not have the cognitive capacity to solve even the simplest of quadratic equations. Other thinkers, like Roger Penrose are cautiously optimistic. In his book *Shadows of the Mind* (Penrose 1996) Penrose outlines four potential positions about the possible neurocomputational basis of consciousness:

Viewpoint A All thinking is computation; feelings of conscious awareness are evoked merely by the carrying out of appropriate computations on the appropriate physical substrate. (This is known as the "Strong Artificial Intelligence (AI)" position).

Viewpoint B Awareness is a feature of the brain's physical action; and whereas any physical action can be simulated computationally, computational simulation cannot by itself evoke awareness. (This is known as the "Weak AI" (Soft AI) position).

Viewpoint C Appropriate physical action of the brain evokes awareness but this physical action cannot be properly simulated computationally. (This is Penrose's personal position).

Viewpoint D Awareness cannot be explained by physical, computational, or any other scientific terms. (This is similar to McGinn's "mysterian" position, although McGinn readily acknowledges that minds greater in capability than those possessed by humankind might, at least in principle, eventually be able to come to understand consciousness and self-awareness in purely scientific terms).

Penrose argues that "we must look beyond the reaches of known sciences if we are ever to find any kind of explanation of the phenomenon of consciousness" (p. 16). According to him, the problem of conscious awareness is a scientific one (not mystical, as in Viewpoint D), even if the appropriate science is not yet at hand. He also allows for graduations between A, B, C, and D. In his earlier book *The Emperor's New Mind* (Penrose 1989) he articulates a position opposite to strong AI by arguing that "there must be an essentially non-algorithmic ingredient in the action of consciousness" and has even gone so far as to speculate that the microtubular structures present in neurons may be important in understanding the physical basis for consciousness.[22]

The above concerns notwithstanding, it is interesting to ask two questions. First, if one were able to produce a conscious artificial being, what tests might one deploy to establish that the new entity is indeed self-aware? And secondly, to what extent should such entities be deserving of moral standing and personhood? That is, what are the philosophical and bioethical consequences of possible future technologies that would be capable of conjuring self-aware beings? These questions are addressed next.

3.20 Consciousness, Artificial Beings and the Turing Test

If we were able to produce a conscious artificial being, what tests might we deploy to establish that the new entity is indeed self-aware? This question is a variant of the well-known "Other Minds Problem", which may be briefly summarized as follows: given that I can only observe the behavior of others and am not granted privileged access to their minds to directly experience their mental events, how can I truly know that any other beings have minds? The problem here is that mere observation of behavior, no matter how sophisticated and complex, does not allow one to infer with complete and absolute certainty that there are mental events associated with this behavior. Some extreme skeptics argue, for instance, that it may actually be the case that all the other people in the universe are in fact unconscious biological automata ("zombies" in the technical parlance used by a number of neurophilosophers). Arguing in this manner that there exist no minds other than one's own and that attempting

[22]Like McGinn, I am not so confident that we will ever find out what causes consciousness to emerge from mere neurons. I expect that in real scientific life, identifying the neuro-computational structures which support conscious existence may be impossible because of overwhelming technical problems related to the complexity of the brain. That is, it might be possible to propose completely accurate brain models without being able to actually do a full test (which might require in vivo brain recordings, for instance). Cosmology has a similar problem with cosmological models such as black holes and baby universes. And, finally, there is even a distant connection of this problem to Godel's Theorem, which tells us that there exist true mathematical statements that cannot be proven to be true.

to establish the existence of other minds is futile is a position known as "solipsism", a position first espoused by the Greek presocratic sophist, Gorgias (c. 483–375 BC).

In the field of Artificial Intelligence (AI) one occasionally offered approach to this problem is the "Turing Test", which may be explained informally as follows. A human judge converses with one human and one computer (or other machine), each of which does its best to appear human. The conversation can be via text communication (for instance via SMS messaging) or by audio, and all participants are isolated from one another to prevent visual cues. If the judge cannot tell the machine from the human, the machine is said to have passed the Turing Test of machine intelligence.

The Turing Test was introduced by Alan Turing in his famous 1950 paper, "Computing Machinery and Intelligence" (French 2000; Turing 1950) and over the following decades it has turned out to be both extremely influential as well as extensively criticized. In 1991 Dr. Hugh Loebner started the annual Loebner Prize competition, offering a prize of US $100,000 and a Gold Medal to the author of the first computer program to pass an unrestricted Turing Test. To date, no computer system has passed. To encourage ongoing efforts, however, each year an annual prize of US $2000 and a bronze medal is awarded to the best effort achieved to date.[23]

It is worth emphasizing that Turing offered his test as an operational means to establish whether a machine could "think". Specifically, his introductory remarks in the paper begin as follows:

> I propose to consider the question, "Can machines think?" This should begin with definitions of the meaning of the terms "machine" and "think". The definitions might be framed so as to reflect so far as possible the normal use of the words, but this attitude is dangerous. If the meaning of the words "machine" and "think" are to be found by examining how they are commonly used it is difficult to escape the conclusion that the meaning and the answer to the question, "Can machines think?" is to be sought in a statistical survey such as a Gallup poll. But this is absurd. Instead of attempting such a definition I shall replace the question by another, which is closely related to it and is expressed in relatively unambiguous words.

As noted earlier, a number of criticisms of the Turing Test have been raised over time. For instance, it is sometimes said to be more a test of a machine's ability to imitate a human being than a test of thinking ability. Another criticism that is sometimes offered is that even if the Turing Test is a test for thinking, this does not necessarily imply that it is a test for consciousness. For a full discussion of this topic, the interested reader is referred to an excellent review by French (2000).

To conclude, while the Turing Test has been offered as a means to consider the question, "Can machines think?", even if it is accepted as a satisfactory test for thinking (despite objections that it is better understood as a test of ability in *imitating* a human being), it does not follow that the Turing Test is a test for sentience or consciousness, a property necessary to being granted moral standing. Discussion on this matter continues below.

---

[23]For the computer enthusiast considering entering the contest, full details are available at http://www.loebner.net/Prizef/loebner-prize.html. In addition, the transcripts for the 2009 bronze medal winner may be reviewed at http://www.worldsbestchatbot.com/Competition_Transcripts.

3.21 Artificial Beings and Moral Standing

Earlier, we asked the question as to the extent to which sentient artificial beings are deserving of moral standing and personhood. Although in the previous section it was made clear that there is no means to establish with absolute certainty that any putative sentient artificial being is in fact sentient, it is reasonable to consider the issues of moral standing and personhood in such beings nonetheless. After all, the "Other Minds Problem" is also applicable to everyday humans and to many advanced species of animals, yet this fact is evidently not an impediment to our considering the issues of moral standing and personhood in these beings.

What then should be our approach? I would propose that an elaboration of the Hughes classification of consciousness and rights provided in Table 3.1 (see Sect. 3.3) be used for just such a purpose. Simply put, any artificial being (robot, cyborg, humanoid, computer etc.) would be classified as to consciousness and rights based upon its behavior and its properties just as humans, animals, embryos and plants are classified in the Hughes classification. I suspect, however, given the likely difficulties of constructing sentient artificial beings, that it will be quite some time before this issue arises in the real world of practical ethics.

3.22 Moral Robots and Robot Ethics

An interesting issue that has arisen among robotics experts in recent years is the possibility of the development of autonomous robots that are programmed with a set of rules regarding their moral behavior. Such a development raises the interesting issue whether moral behavior is a trait unique to humans.[24]

Two examples from the military arena come to mind. Autonomous robots are now entering the theater of war, at least at the conceptual level. Such robots might seek and destroy enemy fortifications and enemy personnel, either operating on the ground or even from the air (perhaps in the form of a small armed helicopter equipped with a video camera and other sensors). The hope among some thinkers is that such "autonomous armed robotic platforms" (as the roboticists like to call such units) will reduce noncombatant casualties and other forms of collateral damage by their unswerving ability to adhere to the "Laws of War" better than even the best trained human combatants. Of course, one can also envision potentially serious problems with such an arrangement, such as problems related to programming errors, ambiguous operational specifications or limitations in the robot's computational power. This is why complex systems such as modern aircraft and nuclear reactors are always designed to allow human "overrides" in their automatic operation.

[24]A related issue concerns moral behavior in higher animals such as the Great Apes. For an interesting discussion on this topic visit http://www.youtube.com/watch?v=GcJxRqTs5nk.

These robots would initially be programmed with a series of specific rules of engagement that might eventually be expanded into a more comprehensive set of guiding principles intended to provide guidance for unanticipated novel situations not covered by a more restricted set of military rules of engagement. In the latter situation, however, the question naturally arises as to which set of guiding principles might be adopted for this purpose.

One might imagine a number of possibilities how such robots might operate. For instance, such robots might have rules not to engage in combat with individuals under a specific height (to avoid harming children), individuals waving a white flag, individuals standing with their arms raised high in the air, or individuals lying prone on the ground with their hands clasped together on their head. Such robots would also be expected to have the means to recognize friendly personnel in the theater of operations, perhaps on the basis of their distinctive uniform or (more reliably) on the basis of an implanted RFID chip that, when interrogated, would respond with a code identifying them as friendly combatants.

A potentially less sinister version of the battlefield combat robot would be a battlefield medic robot that would assist in the extrication of injured soldiers from the battlefield. Like human medics operating in a theater of war, such robots would be armed to defend the patient they are rescuing from the battlefield in order to transport them to a nearby field hospital.

Central to all these scenarios, however, is a set of guiding principles. Thinkers such as Ronald Arkin have proposed that the rules of the 1949 Geneva Convention would be a reasonable starting point for such an initiative.[25] Such rules include principles such as the use of proportionality of force in a conflict, avoiding unnecessary damage to property and noncombatants, the humane treatment of prisoners of war, respectful treatment of corpses, the strict avoidance of torture, and so on.

3.23 Conclusion

In conclusion, developments in fields such as of genetic engineering, neurological technology, artificial intelligence and robotics have forced a reexamination of some basic philosophical problems, such as determining what nonhuman and future parahuman entities are deserving of moral standing and by which criteria. An important consequence of this reexamination was to establish that various degrees of moral standing are due to nonhuman and future parahuman entities on the basis of their cognitive capacities and their ability to experience suffering. These developments have also greatly influenced the growth of the transhumanist movement, as ideas that were once solidly in the realm of science fiction now emerge from laboratories and clinics around the world.

[25]Ronald C. Arkin, *Governing Lethal Behavior: Embedding Ethics in a Hybrid Deliberative/Reactive Robot Architecture.* Undated Technical Report GIT-GVU-07-11 http://www.cc.gatech.edu/ai/robot-lab/online-publications/formalizationv35.pdf.

In the next chapter we address these issues from an additional perspective, that of altering the central nervous system by pharmacological means and by electrical brain stimulation with a view to changing the conditions under which humans live their lives.

Chapter 4
Pharmacologic Enhancement of the Human Condition: Possibilities and Perils

... there is always soma, delicious soma, half a gramme for a half-holiday, a gramme for a week-end, two grammes for a trip to the gorgeous East, three for a dark eternity on the moon...

Aldous Huxley, Brave New World

Abstract This chapter is concerned with the technical, safety and bioethical issues associated with human pharmacologic enhancement. Although neuroscientific advances are now providing valuable treatments for patients with neurological diseases, many of these treatments may also be applied to "normal" individuals free of disease as a means of achieving "quality of life" improvements, an enterprise sometimes called "cosmetic neurology". Although such enhancements might variously make one happier, more serene, more intelligent, better at mental arithmetic, more personable, and so on, they also raise an important philosophical question—could such interventions lead to a mind-blunting effect that would diminish one's capacity to appreciate life in all its dimensions, making us "unauthentic" humans living a banal, anesthetized existence as uncritical participants in the affairs of the world?

Although the term "human condition" is used in widely different ways by various authors, I use the term here to describe the unique joys and challenges of being human, i.e., to describe the constellation of features of humanity that are innate to humans independent of factors such as race, culture or social class. Core issues related to the human condition include the search for the meaning of life, an awareness of the inevitability that life will end, a quest for purpose, gratification and self-actualization, and a host of other issues of a philosophical or psychological nature.

D. J. Doyle, *What Does it Mean to be Human? Life, Death, Personhood and the Transhumanist Movement*, Anticipation Science 3,
https://doi.org/10.1007/978-3-319-94950-5_4

4.1 Introduction

As we learn more and more about the brain and how it works, and especially how the brain's varied receptors and pathways are affected by different classes of pharmaceutical agents, a new age of possibility looms on the horizon. In a mere few decades, continuing advances in the neurosciences will likely come to offer us humans the prospect of safely enjoying a "mood optimized" synthetic existence—an artificial paradise, so to speak—made possible through recent and future advances in neuropharmacology. Such advances may—at least in theory—allow discontented individuals (or even completely normal people) to be pharmacologically liberated from a good number of the negative aspects of the human condition, but—and again, at least in theory—without the many destructive effects of many of today's mood-altering drugs.

Of course, neuropharmacologic enhancement naturally raises specific questions in the context of what it means to be human, although these questions are arguably not substantially different from other potential forms of human enhancement, such as enhancement by way of computer-based neural prostheses, or even by means of manipulating the body's genetic machinery to produce a more favorable gene expression profile. One central question here is this: "To what extent, if any, do enhancement techniques of this kind change our humanness?" This question raises the related question, "What does it mean to be human?", which is, of course, a central theme of this work.

Additionally, I would like to initiate a dialog on two other questions related to the possibilities of neuropharmacologic enhancement: "What might be the anticipated social effects of neuropharmacologic enhancement?" and "Even if such enhancement might be realized without social or personal detriment, should we partake?" Later in the chapter I propose that the while the anticipated social effects of neuropharmacologic enhancement might well be "improved personal and social harmony", the question as to whether we should partake raises important philosophical concerns about the implicit valuation of always feeling 'happy' at the expense of what might be considered more authentically human emotions.[1]

4.2 Might Pharmacological Interventions Help Humanity?

Let us begin by taking a look at some recent developments in pharmacology. To start, it would be prudent to remember the many people whose lives have been totally ruined by drugs. Drugs like methamphetamine ("crystal meth"), cocaine, and heroin, and even prescription drugs like Oxycontin, have at times devastated the social fabric of

[1]This question is itself a philosophically problematic issue, of course—who determines the dividing line between what is "authentic" and what is "inauthentic", and based on to what criteria?

entire neighborhoods,[2] with enormous social costs (Mosher and Yanagisako 1991; Drucker 1999; Hall and Lynskey 2009; Elliott 2004).[3]

On the other hand, a great many drugs have highly desirable clinical properties: antibiotics that cure pneumonia or meningitis, anticancer drugs that melt away malignant tumors, or anesthetic drugs that make modern surgery possible. These drugs are all given for clear clinical purposes.

Complicating this dichotomy, however, is the fact that pharmacologists are now working on developing new drugs that might be best classified as being "cosmetic" in nature (Cakic 2009; Lynch and Gall 2006; Normann and Berger 2008; Morris 2008; Lanni et al. 2008). These drugs would not necessarily be given to treat a specific pathological condition like narcolepsy, epilepsy or Alzheimer's disease (although some of these new drugs would indeed help in such conditions), but instead would be given to "normal" individuals lacking any pathological condition or disease. Such drugs would be given with a view to providing a competitive advantage or other enhancement, such as improving one's memory, augmenting a person's mental agility, or making them more sociable.

Other drugs in early development are expected to provide mood enhancement or even actual euphoria in normal people, but (hopefully) without the addictive or destructive properties of older generation "recreational" drugs. As an example, as the neurochemical correlates of romantic love are slowly being unraveled, it may ultimately be possible to provide a similar mental state by pure pharmacologic means (Debiec 2007; Marazziti et al. 1999; Zeki 2007).

In Peter D. Kramer's book, *Listening to Prozac*, (Kramer 1993) the author suggests that Prozac (fluoxetine) is a drug that can provide enhanced peace and contentment to psychologically normal individuals who wish to favorably change their personality to one that is more socially rewarded. Thus a psychologically normal but melancholic individual might take Prozac to become more socially engaging or to develop more self-confidence. Kramer calls this "cosmetic pharmacology"—the controlled use of psychoactive medications by psychologically normal individuals for the purpose of psychological enhancement rather than treating any identifiable pathological state. Other individuals have called this exciting new field of study "cosmetic neurology" (Chatterjee 2004).

Kramer compares cosmetic pharmacology to cosmetic surgery—the exact same clinical intervention can be *therapeutic* (as in a rhinoplasty procedure to fix a nasal

[2]For a poignant account of how crack cocaine destroyed a Boston neighborhood see http://www.boston.com/news/local/massachusetts/articles/2011/04/10/boston_neighborhood_recalls_devastation_of_crack/. Even better, read the 1997 book, *The Corner: A Year in the Life of an Inner-City Neighborhood* written by Baltimore Sun reporter David Simon and former Baltimore homicide detective Ed Burns. For a more recent (2013) report on the drug-related devastation in some Mexican neighborhoods, the interested reader is directed to a report published by *The Economist*: http://www.economist.com/blogs/americasview/2013/06/violent-crime-mexico.

[3]Drucker (1999) notes that in the USA the period 1979–1996 saw a tenfold increase in imprisonment for drug charges. In 1999, USA "drug enforcement activities constituted 67% of the $16 billion Federal drug budget and more than $20 billion per year in state and local enforcement expenditures, compared with $7.6 billion for treatment, prevention, and research".

obstruction that makes breathing difficult) or *cosmetic* (as in a rhinoplasty to "fix" an ugly nose that nevertheless is perfectly adequate for its intended physiological purpose).

For a variety of reasons, it is possible that in some countries any future "cosmetic drugs" might require a prescription, even if their safety was well-established. For example, in the USA a great many drugs that are available without prescription elsewhere in the world (because of their well-established safety record) nevertheless require a prescription in the USA. One interesting consequence of this is that, in a way, it may end up putting physicians in the unusual role of being a gatekeeper to a satisfying and happy life for psychologically normal individuals. This would be a rather significant change in the usual role of the physician. Additional possible concerns are the likelihood of an illegal market for these drugs (and the need for police action to deal with this problem) as well as demands by the populace that the drugs be made available without prescription, especially if their safety profile proved to be exemplary.

A number of companies are already hard at work developing a variety of other drugs to enhance the human condition. RespireRx Pharmaceuticals Inc. (formerly Cortex Pharmaceuticals, Inc.) is investigating a class of compounds called AMPA receptor modulators, which enhance the glutamate-based transmission between neurons (Ward et al. 2010). Memory Pharmaceuticals, now acquired by Roche and co-founded by Nobel laureate Eric Kandel, is working on an investigational drug called MEM 1003, "an L-type calcium channel modulator", for the treatment of Alzheimer's disease (Rose et al. 2007). Another company, Targacept (now merged with Catalyst Biosciences), is hoping that drugs acting on neuronal nicotinic receptors will lead them to riches and fame (Bencherif and Lippiello 2010). As basic research continues to yield dividends into the understanding of the brain's chemistry, there is little doubt that more intriguing pharmaceutical developments are in our future.

It is not as if researchers are deliberately searching for drugs that might be useful in normal individuals. Rather, the cognitive enhancing properties of these drugs are being discovered serendipitously in research on neurological diseases such as epilepsy, dementia and narcolepsy. As an example, studies in laboratory rats (Goldstein and Davis 1990) and preliminary studies in humans (Aichner et al. 2002) suggest that small doses of amphetamines may be able to help stroke patients undergoing physical therapy to relearn motor skills like tying shoes, buttoning shirts, and using utensils, since it seems to promote neuroplasticity, the process whereby the brain forms new synaptic connections or strengthens existing ones. Although the value of amphetamines in this setting is a matter of some dispute (Walker-Batson et al. 2001), there appears to be little doubt that some approaches to brain enhancement are likely to be spectacularly successful in the long-run.

Such observations make one wonder whether or not the day may come when healthy individuals will take an amphetamine tablet before a golf lesson, knowing that it may enhance the neuronal rewiring that underlies a perfect golf swing, or whether the day will come that surgeons in training will use similar neuropharmacologic tricks to help master the art of constructing the perfect vascular anastamosis. Consider the following examples.

Scenario One: **Movement Enhancement**

Consider that future pharmacologic or even surgical interventions might result in higher performance in conducting gross and fine motor skills. While such interventions might be considered to be unethical for professional athletes or even professional musicians (perhaps by virtue of arguments about an unfair advantage being present), fewer individuals would be expected to object to surgeons improving their psychomotor skills so that complex and risky operations can be completed with greater skill and less morbidity and mortality.

Scenario Two: **Creativity Enhancement**

The question as to whether drugs might be successfully used as a means to enhance creativity has been debated for some time. Samuel Taylor Coleridge (1772–1834) likely wrote his famous poem *Kubla Khan* under the influence of an opium dream.[4] Marijuana was publicly used by Jazz musicians in the 1920s before it became illegal. And, as we discuss later, the famous mathematician Paul Erdös used the drug Benzedrine as a cognitive energizer when tackling unsolved problems in mathematics.

Scenario Three: **Performance Enhancement in Emergencies**

Research suggests that pilots may perform complex tasks better in simulated emergencies when they take donepezil, a drug normally used to treat dementia (Yesavage et al. 2002). Donepezil, an acetylcholinesterase inhibitor, acts by decreasing the activity of acetylcholinesterase, the enzyme which breaks down acetylcholine into choline and acetyl-CoA. By reducing the breakdown of acetylcholine, an increase in the level of brain acetylcholine occurs.

Similarly, Caldwell et al. (2003) reviewed the use of dextroamphetamine (Dexedrine®) as a fatigue countermeasure for use in sleep-deprived military pilots. Their review showed it to be "effective for maintaining flight skills, psychological mood, and physiological activation" (as measured via electroencephalography). Of special importance, the benefits of the drug were not offset by marked disruptions in recovery sleep, although sleep was lighter for several hours following drug administration. The authors concluded that dextroamphetamine "is a viable remedy for fatigue in aviation sustained operations", although "not a substitute for proper crew-rest scheduling because there is no replacement for adequate restful sleep".

Such findings raise the issue whether pilots might occasionally be expected (or even required) to take performance enhancing medications, especially in war time or in the setting of a national disaster.

---

[4]Although *Kubla Khan* was apparently written under the influence of opium, the degree to which he experimented with the drug as a creative enhancement remains unclear. Although Coleridge kept his opium addiction as hidden as possible, it became public knowledge with the 1822 publication of *Confessions of an English Opium Eater* by his friend Thomas de Quincey.

4.3 Some Pharmacophilosophical Questions

The future possibilities discussed above, even if implemented with perfect safety, raise interesting ethical, legal and social questions (Dees 2004). In the case of drugs that enhance a person's mental agility, should such drugs be forbidden when writing university entrance examinations or competing in a chess tournament? What about a mediocre scientist who uses performance enhancing drugs to become a great scientist as a result of enhanced mental skills? This latter idea is not quite as absurd as it might seem to some individuals. As mentioned earlier, the famous mathematician Paul Erdös (1913–1996), did almost all of his highly acclaimed work under the influence of Benzedrine, an early member of the amphetamine family. Or consider the following description of a writer who experimented with Adderall, another amphetamine[5]:

> As an experiment, I decided to take Adderall for a week. The results were miraculous. On a recent Tuesday, after whipping my brother in two out of three games of ping pong — a triumph that has occurred exactly once before in the history of our rivalry — I proceeded to best my previous high score by almost 10 percent in the online anagrams game that has been my recent procrastination tool of choice. Then I sat down and read 175 pages of Stephen Jay Gould's impenetrably dense book ***The Structure of Evolutionary Theory***. It was like I'd been bitten by a radioactive spider.

Of course, Adderall, being an amphetamine, is a potentially dangerous drug, although not so much as methamphetamine, which continues to cause wholesale devastation in some regions of the world (Buxton and Dove 2008).[6] I would also remind the reader that in many countries it is a felony to possess Adderall without a prescription. In fact, in the USA selling and distributing Adderall is a felony punishable by a $10,000 fine or a prison term up to 45 years.[7]

But what if a drug were produced that was like Adderall, with all the beneficial effects described above, but without any deleterious or negative consequences? Many would view such a development as positive. For instance, the Dalai Lama, referring to a method of deep brain stimulation in a talk delivered to the Society for Neuroscience Congress in November 2005, stated: "If it was possible to become free of negative emotions by a riskless implementation of an electrode—without impairing intelligence and the critical mind—I would be the first patient".

The US Declaration of Independence contains the famous words: "We hold these truths to be self-evident, that all men are created equal, that they are endowed by their Creator with certain unalienable Rights, that among these are Life, Liberty and

---

[5]http://www.slate.com/articles/health_and_science/medical_examiner/2005/05/the_adderall_me.html.

[6]Buxton and Dove (2008) state that "nearly 25 million people worldwide are estimated to have used amphetamine and methamphetamine in the past 12 months" and note that "crystal meth is more likely to cause dependence than other forms of methamphetamine".

[7]Dolce N. Adderall use increases, doctors warn of consequences. The Daily Orange. March 7, 2010. Accessed from http://www.dailyorange.com/2.8656/adderall-use-increases-doctors-warn-of-consequences-1.1249380.

the pursuit of Happiness". Surely these words should serve to reinforce the words of the Dalai Lama in American minds.[8]

It is likely that there are a large number of individuals who, despite having no signs of psychopathology, would benefit from safe pharmacological enhancement of their life using drugs likely to be developed in the not-too-distant future. These would not be drugs that are intoxicating or that produce a "high", but rather drugs that would simply make life more pleasant and meaningful. They would be 'happy' pills. And as the US Declaration of Independence appears to tell us, happiness or 'flourishing' is a legitimate life aspiration. However, might not the pursuit of happiness by completely artificial means be fraught with potential difficulties?

One concern about the notion of the pharmacological enhancement of mood, human performance and the like is that it could ultimately (at least hypothetically) lead to unsatisfactory or undesirable outcomes even if found to be biologically safe. Consider a neurochemical agent or other intervention that would bring on the feelings of joy, contentment and romantic love that are so often experienced in newly-wed couples (Bartels and Zeki 2000, 2004). Might chronic use of these or similar agents of happiness produce undesirable permanent brain alterations the way that some currently employed recreational drugs do (Clemens et al. 2004, 2005; Melega et al. 2008; Krasnova et al. 2010)? Or could it just be that the negative aspects actually turn out to be small relative to the benefits, at least in some cases? Such questions cannot be definitively answered at this time, and will remain unanswered until our understanding of neuropharmacology is vastly improved.

Consider the example of clinical anesthesia. In the story of the human conquest over pain, clinicians had to deal with naysayers who objected to using anesthesia for surgery and childbirth (Farr 1980; Swanson 2005). Some individuals even postulate a spiritual or moral value to suffering, claiming that it ennobles them, opening them to new avenues of beauty and splendor and better permitting them to embrace God (Ellis 2002; Carnevale 2009).[9] Might objections to the safe pharmacological enhancement of mood eventually be viewed to be as misguided as early objections against the safe relief of pain and suffering by anesthetic drugs?

Finally, as mentioned earlier, one should consider the judgment implicit in being in a state of perpetual contentment and happiness at the expense of what might be "authentic" human emotions. This last issue is discussed in Sect. 4.7.

---

[8]Note that the phrase "the pursuit of happiness", while often taken to have hedonistic overtones in contemporary parlance, meant something more akin to "the exercise of flourishing" in the language of the time (1776).

[9]On this matter, W. Somerset Maugham wrote: "I knew that suffering did not ennoble; it degraded. It made people selfish, mean, petty, and suspicious. It absorbed them in small things. It did not make them more than human beings; it made them less..." Quote in Phillips D. Z. The Problem of Evil, in *Reason and Religion*, ed. Brown S. C., London: Cornell University Press, 1977, p. 114.

4.4 Imagining a World Without Negative Emotions

Consider for a moment what a world without negative emotions would be like. Imagine a world without hate, anger, resentment, and hostility. Arguably, one possible result would be improved social harmony. Perhaps there would be more stable family units, maybe even less conflict and war.[10]

In the past in a number of "primitive" cultures, marijuana, psilocybin and other drugs played an important role in community life (Nencini 2002). For instance, according to the Dominican friar Diego Durán in his book *The History of the Indies of New Spain* (published c. 1581), hallucinogenic mushrooms were eaten in festivities celebrating the occasion of the accession to the throne of Aztec emperor Moctezuma II in 1502. Similarly, mescaline (in the form of peyote) has been used for millennia by Native Americans in Mexico.

One important goal of ingesting these drugs was to induce a religious or spiritual experience that extended beyond normal perception and served to bring social healing and community harmony (Nencini and Grant 2010). For example, Aldous Huxley's 1954 book, *The Doors of Perception* describes the use of cactus peyote, a psychedelic drug which has been used in American and Mexican Indian religious ceremonies for thousands of years (Huxley 1954).

Similarly, in his 1902 classic book *The Varieties of Religious Experience*, William James suggests that there may be a neurological (and ultimately neurochemical) basis for spiritual experience and religious belief. A century later, this notion has developed into the new academic discipline of neurotheology, the study of correlations of neural phenomena (both natural and pathological) with subjective experiences of religiosity and spirituality.[11] (Related to this is a similar quest to understand the brain mechanisms underlying moral or ethical decision making, a field sometimes known as "neuroethics").

Consider the words of Devinsky and Lai (2008) writing on the neurobiology of spirituality and religion in patients with temporal lobe epilepsy:

> Neocortical areas also may be involved, suggested by the presence of visual and auditory hallucinations, complex ideation during many religious experiences, and the large expanse of temporal neocortex. In contrast to the role of the temporal lobe in evoking religious experiences, alterations in frontal functions may contribute to increased religious interests as a personality trait. The two main forms of religious experience, the ongoing belief pattern and set of convictions (the religion of the everyday man) versus the ecstatic religious experience, may be predominantly localized to the frontal and temporal regions, respectively, of the right hemisphere.

---

[10]This may be more a psychoneurological question than a philosophical one, given that the study of the underpinnings of human emotions is a field of active scientific research. For instance, the journal *Emotion* (http://www.apa.org/journals/emo/) is only one of a number of scholarly journals dealing with the scientific study of emotion.

[11]Neurotheology has potential philosophical implications for addressing the mind-body problem. For example, atheists who subscribe to a materialist conception of mind may argue for a mechanistic understanding of religious experiences to support arguments that religious belief is a mere artifact of brain function.

In a similar manner, Muramoto, a clinical neurologist with a long-standing interest in religion, has hypothesized that the medial prefrontal cortex is important in understanding human religious activity at the neuronal level (Muramoto 2004). He writes that under this hypothesis "optimal functions of the medial prefrontal cortex, such as error detection, compliance to social norms, self-reflection, and theory of mind, are a key prerequisite to the maintenance of integrated religious activity" and further suggests that hyperreligiosity may be the result of "hyperfunction of the medial prefrontal cortex". For further discussion on this fascinating topic, a recent review by Kapogiannis et al. (2009) is recommended.

Although increasing community religiosity and spirituality by pharmacological means might conceivably lead to improved social harmony, it is interesting to explore whether pharmacological approaches might lead to social harmony by means independent of an effect on religiosity and spirituality. Might drugs eventually become available that will increase human compassion and empathy? The recreational drug 3,4-methylenedi-oxymethamphetamine (MDMA, ecstasy, "hug drug") is well-known to increase empathy in those who consume it (Pham and Puzantian 2001). Unfortunately, developing evidence that the drug is neurotoxic with chronic use (as a result of serotonin depletion leading to subtle cognitive and memory impairment) suggests that its ultimate role may be only as a stepping stone in an effort to produce a safer drug with similar effects (Gouzoulis-Mayfrank and Daumann 2006).

In a similar vein, it might be expected that drugs that reduce aggression in humans would also contribute to social harmony. For example, some antiepileptic drugs may reduce aggression in individuals exhibiting neuronal hyper-excitability mediated aggressive behavior (Huband et al. 2010). Other studies suggest that SSRI drugs like Prozac may help control aggressive or violent behavior. In a canine study, fluoxetine (Prozac) was found to be effective in managing dominance aggression(Dodman et al. 1996) Similar results have been found in studies of male lizards (*Anolis carolinensis*) (Deckel 1996).

In humans and in animals, low levels of brain serotonin have been associated with a propensity to impulsive behavior and aggression (Ferris 1996; Haller and Kruk 2006; De Boer et al. 2009). In addition, a number of studies have demonstrated a solid link between testosterone and aggression (Archer 1991, 2006; Olweus et al. 1988), suggesting that drugs that block the effect of testosterone in the brain might be useful as a means to pacify some aggressive individuals. Finally, and perhaps most importantly, there is increasing scientific evidence that the neurochemicals oxytocin and vasopressin can favorably influence social behavior and social cognition (Insel 1997, 2010; Carter and Altemus 1997; Veenema and Neumann 2008; MEYERLINDENBERG 2008; Donaldson and Young 2008; Heinrichs et al. 2009).

Of the above possibilities for neurochemical intervention to influence the human condition, SSRI drugs have received special attention by neuropharmacologists. As an example, consider the following blinded study using normal human volunteers. Knutson et al. (1998) gave subjects either 20 mg/day of paroxetine or placebo for 4 weeks. Relative to placebo, those getting the paroxetine displayed less hostility and a decrease in negative mood (affect). In addition, the drug was found to increase measures of social affiliation. The study authors concluded that "central serotonergic function may modulate a dimension of normal personality characterized by reduced negative affective experience and increased affiliative behavior".

4.5 Brain Enhancement Via Electrical Brain Stimulation

Dr. José Manuel Rodriguez Delgado

Although pharmacologic manipulation of the brain to achieve enhancements in mood or performance appears to be something that we can increasingly expect in the future, it should be pointed out that similar alterations in brain function might also be achieved using electrical methods of brain stimulation.

One of the earliest pioneers in this field was Dr. José Manuel Rodriguez Delgado (1915–2011). In experiments with cats and later with monkeys and humans, Delgado explored the role of electrical stimulation via surgically implanted electrodes as a means to influence behavior.

In one particularly spectacular example, Delgado surgically implanted electrodes in the aggression center of the brain of a belligerent bull. This brain area could be stimulated electronically via remote control. In an interesting public demonstration that was widely reported in the media, Delgado stepped into a bull ring with the bull. When the bull began to charge at Delgado, he pressed a remote control button that instantly stopped the bull from charging. His book, *Physical Control of the Mind: Toward a Psychocivilized Society* (1969), Delgado explored the various possibilities that might be developed with further research in this technology. While the notion of physical control of the mind naturally conjures up an image of people with implanted electrodes responding to a master who controls their thoughts and actions (as appeared to be the case with the example of Delgado's bull), Delgado goes to lengths to attempt to dispel such notions, emphasizing that the value in these methods is primarily to obtain an improved understanding of how the human brain functions, not to construct robots out of humans. That being said, the final chapter of Delgado's book outlines a plan for "conquering the human mind" through increased neurobehavioral research that would have critical implications for our understanding of personhood. This is because altering the temperament and outlook of a person via pharmacological, neurobehavioural and other methods of intervention changes their very nature. They are no longer their own person, but have been shaped, altered and edited. While Delgado suggests that when a person's behaviour conflicts with society so seriously to result in, say, incarceration, such intervention could be justified, it should still be clear that the post-intervention individual is not the same person—thus the implication for our understanding of what it means to be a person. Note that while such interventions may completely change an individual's nature, this fact in itself does not imply that same individual necessarily undergoes a loss of personhood and the moral standing that accompanies personhood—these individuals simply become different persons rather than nonpersons.

One intriguing field related to Delgado's research is that of electrical stimulation, via implanted electrodes, of the brain's "pleasure center", usually (but not always) taken to be the nucleus accumbens (Ikemoto and Panksepp 1999). The nucleus accumbens is a dopaminergic structure, part of the limbic system, believed to play an important role in the "high" obtained from certain recreational drugs, as well as to play a part in sexual arousal (Ikemoto and Panksepp 1999). Stimulation of the nucleus accumbens has been studied in humans in efforts to treat severe obsessive compulsive disorder (Huff et al. 2010) severe depression and anxiety (Bewernick et al. 2010) and even drug addiction (Stelten et al. 2008).

In one case of a 47-year-old woman with "chronic treatment-refractory obsessive-compulsive disorder, nicotine dependence, and obesity," treatment with deep brain stimulation of the nucleus accumbens for obsessive-compulsive disorder resulted in "unintended, effortless, and simultaneous smoking cessation and weight loss" (Mantione et al. 2010). The authors suggested that "deep brain stimulation of the nucleus accumbens could be a possible treatment of patients with a dependency not responding to currently available treatments". Such a clinical report has explicit philosophical relevance in so far as it exemplifies the tight coupling between the operation of the brain and the operation of the mind. This has obvious implications for the notions of "self" and "identity", given that the operation of the mind (which embodies self and identity) can be so fully changed via physical interventions.

Finally, it is an interesting aside to note that the nucleus accumbens is now believed to be an important neurophysiological structure central to the placebo effect (Scott et al. 2007; Zubieta and Stohler 2009).

While some critics, especially in Delgado's time, expressed concerns that such electronic implants might be used in humans to reduce individuality and control behavior that deviated from the norm,[12] experience in the last few decades since his pioneering work has gone in a rather different direction, with other recent application of this technology directed at the management of conditions such as Parkinson's disease (Wood et al. 2010), Tourette syndrome (Porta et al. 2009), the management of chronic pain (Levy et al. 2010), epilepsy (Kahane and Depaulis 2010) and cervical dystonia (Huh et al. 2010).

Transcranial Direct Current Stimulation

The brain stimulation methods mentioned so far involve implanted electrodes placed via a neurosurgical procedure. However, recent research suggests that some forms of cognitive enhancement might be achieved much more easily, using direct current applied via simple scalp electrodes. Known as Transcranial Direct Current Stimulation (tDCS), it involves supplying a weak electrical current (typically 2 mA or less), usually for a period of 5–30 min. A variety of electrode placement options exist, depending on the intended effect (Thair et al. 2017; Stagg and Nitsche 2011).

[12]Remember that in Delgado's time the horrors of Nazism and Stalinism were still strong in many people's minds.

Because studies on tDCS suggest that it can accelerate learning and boost task performance as well as improve mood and social cognition (Parasuraman and McKinley 2014) it has been the subject of investigation by both professionals and hobbyists (visit www.diytdcs.com for a sample hobbyist hangout). However, the use of tDCS by hobbyists worries some individuals, who are concerned that it is too soon be sure that tDCS is as safe as it appears to be, that it should only be administered by specially trained individuals who use only established and tested protocols, and that it might be administered without full informed consent, or even administered to children by parents wishing to provide their school-aged child with a cognitive advantage (Chi et al. 2010; Lifshitz-Ben-Basat and Mashal 2018). Additionally, should tDCS be proven to be as effective as methods involving implanted electrodes (so far the evidence appears to suggest otherwise) the ethical issues discussed later related to brain enhancement via implanted electrodes would also apply.

The pharmacologic and brain stimulation studies outlined here suggest that the possibility of improving the human condition and increasing social harmony by widely-applied pharmacologic means (and/or by electrical means) is no longer simply a wild pipe dream. One can even imagine controlled alteration of a nation's water supply to achieve these goals in a manner similar to the fluoridation of water to reduce dental caries. The question, however, thus appears to be not "can we?", but rather "should we?" In the next sections we will explore this issue from an existentialist perspective.

4.6 But Should We Partake?

Supposing that such a pharmacologic paradise as we have been discussing might actually be possible, that a drug or series of drugs or other intervention might make our lives more joyous and fulfilled, it is still not immediately clear that it would be prudent to partake, even if apparently safe.

Of course, any apparent safety might still be an illusion. For instance, when heroin was first synthesized by the Bayer Company in 1897 it was not believed to be addictive! (Metzger 1998). (It was first sold as a cough suppressant). But, given the great advances in drug safety that have taken place in recent decades, perhaps safety issues should not be our primary concern.

Instead, I would like to raise a rather different concern. Let us focus instead on the impact on personhood and the human existential condition. Could it be that such mind-altering interventions might lead to a mind blunting effect that would diminish one's capacity to appreciate life in all its dimensions? Would not getting joy and fulfillment via a pill risk making us "unauthentic" humans living a banal, anesthetized existence as uncritical participants in the affairs of the world? In this scenario, injustices would not bother us, moral outrage would fade away, and shame at wrong-doing would simply no longer bother us.

The reader may be reminded of Aldous Huxley's classic dystopian novel ***Brave New World***, where the drug soma was used ubiquitously for mood enhancement.

Similarly, Michael J. Sandel's book *The Case Against Perfection: Ethics in the Age of Genetic Engineering* (Sandel 2007) offers similar concerns. Whether it is called "Paradise Engineering", "false happiness" or "sugar-coated pseudo-realism", thinkers such as Sandel argue that it would be best to decline participation in this future until its consequences become more obvious. Let us explore these concerns further, drawing on some existentialist notions.

4.7 What Does It Mean to Be an "Authentic" Human Being?

Since the time that Socrates commented that "the unexamined life is not worth living" philosophers have examined what it means to be authentic, albeit in a variety of culturally contingent guises. The word "authentic" may be defined as "conforming to fact and therefore worthy of belief", or "the quality of being genuine or not corrupted from the original", or "of undisputed credibility".[13] Among existentialist philosophers the concept of authenticity has been particularly well developed, often with reference to the extent to which one is true to one's own character despite the pressures and forces of the world we inhabit.

Søren Kierkegaard (1813–1855) was a Danish philosopher who is widely considered to be the father of existentialist philosophy. Kierkegaard wrote widely on issues of personal choice and commitment, especially in the context of religion, where he is best known for his concept of "leap of faith", an act of accepting the unprovable.

Martin Heidegger (1889–1976), whose analyses of *Dasein* in his 1927 magnum opus, *Sein und Zeit* paved the way for subsequent existentialist philosophies, emphasized that to lead an "authentic" life, one must exercise one's personal freedom to establish a meaningful existence for oneself and take responsibility for the consequences of our choices, rather than follow a path of self-deception where one simply follows orders, accepts conformity, and is concerned only with the here and now.

One of Heidegger's most important concepts in relation to authenticity is "Being-toward-death". To Heidegger, Being-toward-death forges a truly authentic being, as death belongs only to the one who experiences it; it cannot be shared. Being-toward-death acknowledges one's finitude, and therefore the facticity of being-in-the-world, i.e. what it means to "live in the now".

Heidegger points out that leading an authentic life is not always easy, as the process entails uncertainty and anxiety, and requires considerable courage. (Undoubtedly, Heidegger would have rejected the idea of taking "happy pills" as a measure against existential angst!).

Another existentialist philosopher concerned with authenticity was Jean Paul Sartre (1905–1980), who shared a "common philosophical ancestry" with Heidegger in that both were concerned with the problem of existential worry in a meaningless, even absurd, world. Sartre's novels repeatedly describe characters and anti-heroes

[13] http://en.wiktionary.org/wiki/authenticity.

who succumb to external pressures in order to have a more comfortable existence. They do this even though in succumbing to these pressures their actions are contrary to their own moral and aesthetic principles. (For example, one might avoid fighting against oppressive political or social circumstances by saying "I am not free to risk my life in any such struggle—I have a family to support"). Sartre calls this "bad faith", a psychological phenomenon whereby one denies one's freedom in order to avoid taking action with respect to one's circumstances, of seeing oneself not as a free person but as a powerless object.

The classic example of bad faith that Sartre provides is that of a restaurant waiter whose exaggerated behaviours (pleasing but scripted tone of voice, perfect stance, etc.) are all aimed at satisfying his clients. He acts as an automaton whose essence is to be a waiter rather than a human being, adopting false values and ignoring their innate freedom.

(Since reading of Sartre's waiter many years ago, I have come to recognize an interesting parallel in my own interactions as an anesthesiologist dealing with anxious patients who are about to undergo surgery. Rather like Sartre's waiter I too frequently utilize reassuring but subtly scripted phrases and behavior to make my patients more comfortable as they are about to undergo what is very often one of the most terrifying experiences of their life. In addition, at least until the patient is rendered unconscious, I attempt to provide an atmosphere of extreme professionalism that, despite being partly scripted and staged, is nevertheless very reassuring to the patient. But is this not also like being an automaton or robot, one whose essence is to act as best as possible like a caring and compassionate physician? To the extent that such actions are played out using an informal script optimized over time for maximum patient psychological comfort, instead of being based on one's natural personality, is this not also a form of "bad faith"?). In the final analysis, however, I see an important weakness in this analogy: the physician acting to obtain the best clinical outcomes and providing the best possible care to his or her patients can hardly be said to be adopting false values and acting inauthentically.[14]

A number of thinkers have discussed the issue of authenticity in the context of psychopharmacology and pharmacological enhancement. For instance, Aldous Huxley's book *Brave New World* and a number of lesser known literary works address this issue in various ways. In the case of *Brave New World*, Huxley described a dystopian world where a drug called Soma played a central role. Many years later, Huxley wrote *Island*, a novel with a distinctly utopian theme, also containing a psychopharmacology thread. But while *Brave New World* emphasized drug use for pacification of the populace, *Island* emphasized the potential of psychopharmaceuticals for the more noble purposes of enlightenment, improved learning and enhanced self-knowledge (Schermer 2007).

[14]There is something paradoxical in the physician–patient relationship: on the one hand he/she must always be the objective scientist who—to an extent—objectifies the patient. On the other hand, the physician must also be subjective in the sense that he/she realizes that the object is a human being like the physician themselves. Thus, the physician–patient relationship is both a subject-object relationship and a subject–subject relationship.

More recent thinkers have added to this discussion. Elliott (Elliott 1998a, 2003) expresses an unease concerning the administration of psychopharmacological substances to psychologically normal individuals suffering from existential angst that is more a consequence of the human condition than any mental pathology. Arguing that psychopharmacological enhancement threatens our human dignity and our connection to reality, Elliott raises the concern of living an inauthentic life, a life giving you (for example) a better, more pleasant personality, but one that is not ***your*** personality.

On the other hand, DeGrazia (2005) takes the position that a "lucid" understanding of human nature and human identity, along the clear concepts of enhancement and authenticity "largely neutralizes" such concerns, which DeGrazia holds to be "seductive, but unsound". Bolt (2007) offers a similar line of argumentation, although he argues that DeGrazia's conception of authenticity is inadequate to the task, as "his conception of authenticity does not explain the reason why we find inauthentic values objectionable". In fact, from a philosophical point of view, as we have seen, both "enhancement" and "authenticity" are value-laden concepts subject to considerable debate.

Schermer (2007) takes the position that we should not "get stuck in an opposition of dystopian and utopian views", such as those presented by Huxley, but should instead focus on real-world issues: the "evaluation and governance of enhancing psychopharmacological substances in democratic, pluralistic societies". Thus Schermer would have us employ a scientific, rational, risk-based approach to drug policy rather than the politically-driven process that so often dominates the current discourse on drug policy.

Erler (2011) considered the question whether memory modification technologies "might lead us to live inauthentic lives". Distinguishing between memory enhancement and memory editing (for example, to eradicate painful memories), Erler argues that while memory enhancement does not raise serious authenticity concerns, memory editing does: "first, by threatening its truthfulness, and secondly, by interfering with our disposition to respond in certain ways to some past events".

Kraemer (2011) argues against the sometimes espoused position that psychopharmacological interventions will always lead to diminished emotional authenticity.[15] He argues instead that clinical case studies "reveal that some psychotropic drugs such as antidepressants actually cause people to undergo experiences of authenticity, making them feel 'like themselves' for the first time in their lives". He develops the position that the so-called "authenticity standard" for emotions (requiring that emotions not be the result of artificial interventions, such as from drugs) is not always the best standard, and that other ways of assessing emotions, such as a "rationality standard" or a "coherence standard" (looking for harmonized experiential, behavioral, and physiological responses) may be more appropriate in parsing the ethical debate on psychopharmacological manipulation of emotions.

---

[15]For an examples of this point of view, see Fukuyama's Our *Posthuman Future* (Fukuyama 2002) and Elliot's *The Tyranny of Happiness* (Elliott 1998b).

4.8 Back to Religion

A similar "authenticity" argument has at times been offered to critique the role that religion sometimes plays in influencing how one copes with one's present life circumstances. It is certainly a truism that many individuals find extreme contentment and fulfillment in religion. However, some critics have argued that a religious belief in a glorious afterlife in paradise can serve to diminish concerns about oppression and injustice on earth, with inaction against injustices being justified on the basis that the whole point of life on earth is simply to make it into heaven, not to achieve happiness and fulfillment in this life. For example, in his 1886 book *Beyond Good and Evil* and his 1887 book *On the Genealogy of Morality,* Nietzsche famously argues that Christian morality is a life-negating product of self-deception that reflects an inferior, subservient ("slave") morality. Christianity, he argues, does not encourage individuals to flourish but instead asks believers to suppress human instinct and passion and become passive human beings stripped of those very properties that make them most human.[16]

Although Marx was probably not the first to make the suggestion that "religion is the opiate of the people"[17] he is famously known for his contempt of religion and his proposal to replace belief in God with a materialist philosophy that makes no appeal to the supernatural. It is undeniably true, however, that when man's existence on earth becomes intolerably brutal and nasty, the assurance of an afterlife in paradise can be extremely comforting for some individuals, regardless of whether or not the belief is ultimately justified. The debate as to whether any particular set of religious beliefs is warranted by the evidence is beyond the scope of the present discussion. However, the evidence that a great many individuals act as if afterlife in paradise awaits them is undeniable, as evidenced, for instance, by suicide bombers who readily sacrifice their life for a distorted and pathologic vision of what is right and just.

To the extent that belief in a set of religious principles, misguided or otherwise, can lead to contentment and happiness, it can be argued that religion can be made to play a role in ameliorating the human condition, as also might pharmacologic enhancement or even merely finding oneself in the midst of a joyous relationship with friends and family. Arguably, it is only when adherence to a set of religious beliefs leads to social disorder and personal suffering that they can be seen to be not helping improve the human condition.

What, then are we to make of all the above? Will the pharmacologic future that awaits us ultimately improve the human condition, as did the invention of anesthesia, or does it open more dystopian possibilities of a more sinister nature? My view is that it still too early to tell, and on careful reflection I must concede that given the enormous complexities of the human mind, the question as to the desirability of

---

[16]These comments notwithstanding, it is also a truism that many religious people are selflessly devoted to trying to help improve humanity and its future. Additionally, not all religions believe in an afterlife.

[17]See http://www.quotecounterquote.com/2011/10/religion-is-opium-of-people.html for an interesting, if brief, discussion.

such exciting possibilities is ultimately an empirical question that simply cannot be answered at this time.

That being said, based on what we know from the existing debate, I rather expect that the pharmacologic future that waits will ultimately improve the human condition, at least in a number of settings. And should such options for improvement ultimately take place, they will necessarily cause many individuals to reevaluate what it means to be human.

Chapter 5
Life, Death, and Brain Death

It remains for the doctor and especially the anaesthesiologist, to give a clear and precise definition of "death" and the "moment of death" of a patient who passes away in a state of unconsciousness.

Pope Pius XII (Address to an International Congress of Anesthesiologists (November 24, 1957), in Acta Apostolicae Sedis, 49 (1957), p. 1030).

Abstract This chapter is primarily concerned with the issues associated with defining death and the philosophical debate on where the line between life and death should be drawn. While traditionally death has been defined as the permanent cessation of the heartbeat and respiration, modern clinical developments have forced a reevaluation of the both the concept of death in general and of brain death in particular. Future developments in cryonics and nanotechnology may require even further reappraisal of when a person is really dead. This chapter advances the philosophical position that organismal death is not an event so much as it is a process, and that life and death are not distinct binary states, but entities with degrees and gradations. Finally, it advances the notion that it is helpful to distinguish between the definition of death based on mere cessation of the heart beat and a "more substantial" form of death, such as exists following cremation ("absolutely irreversible death" or "information-theoretic death").

5.1 Introduction

In previous chapters we discussed what it means to be a human and what it means to be a person in a world complicated by astonishing advances in biomedical technology. In both cases, the presumption that we are dealing with a living organism is made, as the dead are considered neither to be humans nor to be persons. In this chapter the moral, ethical and legal problems associated with brain death in both its complete and incomplete forms are presented. It is argued that although the notions of personhood

D. J. Doyle, *What Does it Mean to be Human? Life, Death, Personhood and the Transhumanist Movement*, Anticipation Science 3,
https://doi.org/10.1007/978-3-319-94950-5_5

and personal identity are central to the discussion, the notions of "life" and "death" are surprisingly ambiguous. In this sense this chapter and the next one dealing with cryonics break new philosophical ground in their contribution to the debate on where the line between life and death should be drawn.

5.2 What Is Life?

The exact nature of life has puzzled biologists (and no small number of philosophers) for a great many centuries. "Living organisms are autopoietic systems: self-constructing, self-maintaining, energy-transducing autocatalytic entities" in which information needed to construct the next generation of organisms is stabilized in nucleic acids that replicate within the context of whole cells and work with other developmental resources during the life-cycles of organisms, but they are also "systems capable of evolving by variation and natural selection: self-reproducing entities, whose forms and functions are adapted to their environment and reflect the composition and history of an ecosystem" (Harold 2001).[1]

"What is Life?" is the title of a 1944 book by Erwin Schrödinger, the 1933 winner of the Nobel Prize in physics for his seminal contributions to quantum mechanics.[2] Among other things, Schrödinger's book asks "how can the events in space and time which take place within the spatial boundary of a living organism be accounted for by physics and chemistry?" (p. 3). Based on the observation that all living organisms have the ability to remain in a state of order (decreased entropy) despite the natural tendency for all things to decay into disorder, Schrödinger postulated that all living matter evades decaying towards a thermodynamic equilibrium (as would ordinarily be expected as a consequence of the second law of thermodynamics) by feeding on "negative entropy" (negentropy).[3] This is yet another approach to understanding what is meant by "life."

The above commentary notwithstanding, a number of scientists argue that matters concerning the definition of life are not as nearly clear-cut as they might appear. Cleland and Chyba (2002) note the following:

> There is no broadly accepted definition of "life." Suggested definitions face problems, often in the form of robust counter-examples. Here we use insights from philosophical investigations into language to argue that defining 'life' currently poses a dilemma analogous to that faced by those hoping to define 'water' before the existence of molecular theory. In the absence

[1] Stanford Encyclopedia of Philosophy. http://plato.stanford.edu/entries/life/.

[2] The book can be downloaded at http://csustan.csustan.edu/~tom/life/What-is-Life.pdf.

[3] The second Law of Thermodynamics states that the entropy (degree of disorganization) of a system tends to get larger as time progresses. Living things, however, behave in an opposite manner, organizing their environment for their own biological needs, creating order from disorder. Schrödinger uses the term negentropy to identify this fundamental ability of living systems, an ability to not only avoid entropy production—as dictated by the second law—but to do just the opposite, to increase organization, exporting entropy to keep its own internal entropy low.

> of an analogous theory of the nature of living systems, interminable controversy over the definition of life is inescapable.

Similarly, Oliver and Perry (2006) write:

> Although there have been attempts at a definition of life from many disciplines, none is accepted by all as definitive. Some people believe that it is impossible to define 'life' adequately at the moment. We agree with this point of view on linguistic grounds, examining the different types of definition, the contexts in which they are used and their relative usefulness as aids to arriving at a scientific definition of life.

As a practical example, one issue of continuing controversy in biology is whether or not viruses and prions should be considered to be forms of life.[4] Most biologists argue that these entities should not be considered forms of life, since (for instance) viruses do not metabolize, do not react to stimuli and cannot reproduce without invading a cell and altering its molecular machinery.

A related issue is what constitutes *human* life. This is a matter that has been addressed on and off in earlier sections of this work and is a matter we will continue to be concerned with in the following chapters.

A final issue deals with the question of when human life can be said to have ended. This is an unexpectedly complex issue that is discussed next.

5.3 The Traditional Concept of Death

Biologists frequently define death as the permanent cessation of all biological functions that define a living organism. To clinicians, however, death of the human organism has traditionally been defined more simply as the permanent cessation of the heartbeat and respiration. Although this determination is usually not difficult to make, mistakes occasionally occur. For example, in Victorian times there were stories of individuals mistakenly being declared dead and later 'returning to life' when embalming began. See, for instance, Edgar Allen Poe's horror story "The Premature Burial", published in 1844, as well as Antoine Wiertz's similarly titled 1854 painting produced in homage to Poe's story. In fact, the fear of being buried alive in Victorian times was so prevalent that some individuals requested that following their apparent death that they be buried in a coffin with an external bell mechanism that they could operate from inside their coffin should they ever regain consciousness while buried.

Even today, mistakes in the diagnosis of death occasionally occur. A 2005 article in the New York Times[5] recounts the appalling story of a man (Larry Green) found at the scene of a car accident, "where he was examined by paramedics and declared dead", then "zipped into a black vinyl bag, taken to the morgue and slid into a refrigerated drawer." Only later, when other personnel "noticed what appeared to be

---

[4]A prion is an infectious agent consisting of a misfolded protein and containing no nucleic acids. Prions are responsible for Creutzfeldt-Jakob disease and a number of other infectious conditions.

[5]http://www.nytimes.com/2005/02/13/national/13dead.html.

an in-and-out movement in Mr. Green's chest and abdomen," did it become apparent that the initial diagnosis of death was a regrettable and tragic error, one likely based on an excessively cursory clinical examination conducted at the scene of the accident.

Despite the attractive simplicity of diagnosing death based on absent pulse and respiration, modern developments in clinical resuscitation have forced a reappraisal of this appealing but outdated notion. Today, in hospitals around the world, ventilators, dialysis equipment and drug infusions that sustain the circulation often permit the bodies of critically ill patients to be artificially supported despite severe physiological insults, including death of the brain itself. In addition, the advent of transplantation surgery has provided a strong clinical motivation to allow death to be defined in terms of loss of brain function (brain death) or in terms of a relatively brief period of asystole (cessation of cardiac activity) in some patients (so-called non-heart-beating organ donors). Finally, some thinkers argue that future developments in cryonics and nanotechnology may require further reappraisal of when a person is really dead. These issues are the subject of this and later sections, which present as a central theme the notion that the concept of death is neither crisp nor static but rather has elements of indistinctness and ambiguity, at least at the extreme margins of human existence.

5.4 Brain Death

While the notion of death as permanent cardio-respiratory arrest has been familiar to physicians since biblical times, the notion of brain death is much more recent, having been developed in the 1960s as advances in transplantation medicine and medical technology provided a motivation to allow death to be defined in terms of loss of brain function (Truog and Robinson 2003). This was because the success of transplantation surgery depends critically on the use of viable organs uncompromised by circulatory failure.

In a typical clinical scenario, a patient with a massive brain injury (for example, from a ruptured cerebral aneurysm) is treated in an intensive care unit. While there, the patient's neurological status is periodically assessed and if there is suspicion that the patient is brain dead, a formal evaluation for possible brain death is carried out. If the patient is found to meet brain death criteria, the patient's relatives are often approached regarding possible organ donation. If they agree, cessation of life support measures are carried out after organ retrieval for transplantation has taken place. Cardiopulmonary collapse and classical death usually follows promptly.

A key pathophysiological event in brain death is a severe elevation of intracranial pressure (ICP) from causes such as hemorrhage (bleeding) or edema (swelling). When the ICP rises to approach arterial blood pressure, cerebral perfusion pressure (CPP) approaches zero, cerebral perfusion (blood flow to the brain) stops and brain death ensues. Autopsy findings of the brain depend on the precipitating event but may include total liquification in cases of prolonged artificial life support (sometimes known as "respirator brain"). With primary hemispheric lesions (e.g., large unilateral tumors), transtentorial or tonsilar herniation are often present. A careful neurological examination centering on the assessment of brain-stem reflexes is the basis for the determination of brain death. The diagnosis of brain death is determined primarily on the basis of clinical findings, and depends on investigations such as electroencephalography, Doppler blood flow studies and cerebral angiography only in special cases (vide infra).

5.5 Early Brain Death Criteria

Associated with the concept of brain death is the need for a reliable, structured method of diagnosis. A variety of such criteria have been published. Of these, the original Harvard Criteria (Ad Hoc Committee of the Harvard Medical School to Examine the Definition of Brain Death 1968) are perhaps the best known. They include the following requirements: (a) unresponsivity to intensely noxious stimuli (unresponsive coma), (b) total absence of spontaneous breathing, (c) absence of brainstem and spinal reflexes, (d) absence of postural activity such as decerebration, and (e) a flat electroencephalogram (EEG). Hypothermia and the presence of CNS depressants such as barbiturates and benzodiazepines must also be excluded. Finally, the clinical and EEG findings should be unchanged in a second evaluation at least 24 h later.

More relaxed criteria such as the Minnesota Criteria (Mohandas and Chou 1971) were later proposed. Notably absent from this criteria list are requirements for absent spinal reflexes and a study of electroencephalographic (EEG) activity (electroencephalography being viewed as an optional confirmatory investigation in this instance). The key elements of the Minnesota criteria are: (a) absence of spontaneous movement, (b) absence of spontaneous respiration over a four-minute test period, (c) absence of brain reflexes as evidenced by: fixed dilated pupils; absent gag, corneal and ciliospinal reflexes; absent doll's eye movements; absent response to caloric stimulation, and absent tonic neck reflex, (d) unchanged status for at least twelve hours and (e) responsible pathological process deemed irreparable.

Over time, continuing refinements to brain death criteria have taken place. In 1981 the legal basis for brain death in the USA became established based on an important report by the President's Commission for the Study of Ethical Problems in Medicare and Biomedical and Behavioral Research (1981). Since that time this report has been used as the basis for a great deal of the discussion on the criteria for brain death. Finally, note that Although not explicitly noted in all criteria for brain death, hypothermia, the presence of central nervous system (CNS) depressants such as opiates, ethanol, barbiturates and benzodiazepines, and the presence of severe metabolic abnormalities (such as severe hypoglycemia or severe hyponatremia) must also be excluded.

5.6 Ancillary Tests for Confirming Brain Death

Although the diagnosis of brain death is a clinical one, that is, not based on laboratory tests, from time to time the clinician may seek guidance from a variety of ancillary tests. Perhaps the most common such test is electroencephalography, the expected finding being a "flat" or "isoelectric" electroencephalogram (EEG). As well, confirmation of brainstem death can be facilitated by observing a flat auditory brainstem response (ABR). Absence of cerebral circulation can be demonstrated by four-vessel angiography as well as by radionuclide methods. More recently, absent flow signals on transcranial Doppler ultrasonography has also been suggested as a method to demonstrate the absence of cerebral circulation in brain death.

All these methods suffer from potential technical issues in that they require skilled technical support to make the techniques meaningful and reliable in the clinical environment. In addition, reasonable, experienced clinicians will sometimes disagree about when ancillary tests are appropriate in confirming the diagnosis of brain death.

5.7 Death of the Entire Brain as a Requirement for Brain Death

As noted earlier, as clinical experience evolved, criteria for brain death continued to be refined so that in 1981 the legal basis for brain death in the USA came to be based on a profoundly influential report by the President's Commission for the Study of Ethical Problems in Medicare and Biomedical and Behavioral Research (President's Commission for the Study of Ethical Problems in Medicine and Biomedical and Behavior Research 1981). The commission defined brain death as the "irreversible cessation of all functions of the entire brain, including the brain stem." Operational criteria for diagnosing brain death were not explicitly addressed in the report; this aspect was dealt with by another committee composed of medical experts rather than bioethics experts who published a contemporaneous report in the *Journal of the American Medical Association* (JAMA).

Since that time this report has been used as the basis for much of the discourse on brain death. For instance, influenced by this report, the World Medical Association has written that in declaring brain death "[i]t is essential to determine the irreversible cessation of all functions of the entire brain, including the brain stem".[6]

[6]World Medical Association Declaration on Death, available at http://www.wma.net/en/30publications/10policies/d2/index.html.

5.8 Ignoring the "Death of the Entire Brain" Rule in Declaring Brain Death

While august bodies such as the President's Commission for the Study of Ethical Problems in Medicare and Biomedical and Behavioral Research and the World Medical Association have defined brain death in terms of "irreversible cessation of ALL functions of the ENTIRE brain," in the years since this definition has been widely adopted it has become very clear that many (perhaps most) patients diagnosed with brain death (and who have undergone organ harvesting) do not actually meet this requirement (Halevy and Brody 1993). In particular, many patients diagnosed with brain death still have some residual function in the hypothalamus, a part of the brain that contains a number of important nuclei, some linking the central nervous system to the endocrine system via the pituitary gland (hypophysis). Even more surprisingly, evaluation of hypothalamic function is not part of any brain death protocols in common use in the USA (or, apparently, elsewhere in the world).

To many thinkers, such a state of affairs borders on the incredible. It is, however, true beyond question. But then, how can the global clinical community simply ignore hypothalamic function in declaring brain death? Could it be that many patients declared to be brain dead are not actually fully brain dead? And surely, "not fully brain dead" is not really dead at all. Not surprisingly, inconsistencies of this kind in the declaration of death are troubling to a great many clinicians, who would prefer that any possible logical inconsistencies in the declaration of brain death be avoided, no matter how small or clinically unimportant they may appear to be (Baron et al. 2006). Philosophers, too, are obviously entitled to some degree of intellectual bewilderment on this matter. In addition, so are some individuals who have signed organ donor cards.

5.9 Some Brain Stem Nuclei Are not Tested When Testing for Brain Stem Death

The above inconsistency notwithstanding, testing for brain death requires that irreversible cessation of all functions of the entire brain, including all functions of the brain stem, be present. Consequently, one would expect that all 12 cranial nerves be shown to be nonfunctional in any declaration of brain death. Cranial nerve testing for the purposes of brain death testing has been described by Wijdick (2001): testing according to standard methods require the absence of grimacing or eye opening with deep pressure on both condyles at the level of the temporomandibular joint (testing afferent nerve V and efferent nerve VII), an absent corneal reflex elicited by touching the edge of the cornea (thus testing cranial nerves V and VII), an absent light reflex (thus testing cranial nerves II and III), an absent oculovestibular response toward the side of the cold stimulus provided by ice water (thus testing cranial nerves VIII and III and VI), and an absent cough reflex elicited through the introduction of a suction catheter deep in the patient's trachea (thus testing cranial nerves IX and X). As can be seen, this protocol tests cranial nerves II, III, V, VI, VII, VIII, IX and X. However, not all the cranial nerves are tested when brain death testing is conducted by standard protocols. Specifically, cranial nerves I, XI and XII are not tested. This is because some cranial nerves are simply too difficult to test.

This raises an obvious concern. Brainstem testing of cranial nerves only tests the nerves that are accessible for testing in unconscious individuals. If any of the testable cranial nerves are actually shown to be operational, then brain stem death cannot be declared. But, according to current brain death protocols, brain stem death *can* be declared should one of the untestable cranial nerves (untestable in unconscious patients) actually be operational in reality (even though we have no way of testing the untestable in unconscious patients with current technology).

This is not a mere matter of semantics, as one could imagine a newly developed test for, say, cranial nerve XI that would allow us to test for its function in unconscious individuals. See, for instance, the following case report abstract concerning an apparently functioning cranial nerve XI in a patient declared to be brain dead (Patterson and McShane 1991). This suggests that ancillary tests such as cerebral angiography may on occasion erroneously establish the absence of cerebral circulation or may not have enough power to detect small amounts of cerebral circulation sufficient to keep some of the brainstem nuclei alive.

5.10 The Problem of Spontaneous Movements Despite Brain Death

Both classical criteria such as the Harvard criteria as well as more recent criteria for the clinical diagnosis of brain death require that the patient display no spontaneous movements. For instance, Shemie (2006) writes: "Deep unresponsive coma implies a lack of spontaneous movements as well as an absence of movement originating in the CNS, such as cranial nerve function, CNS-mediated motor response to pain in any distribution, seizures, decorticate and decerebrate responses. Spinal reflexes or motor responses confined to spinal distribution may persist." Note that some movement in response to noxious stimuli is permitted, as long as these movements are spinally mediated and do not originate from the brain itself.

CASE REPORT

A 56 year old woman was admitted to our hospital following a fall in which she sustained an acute subdural haematoma. She failed to respond to treatment despite surgical intervention and active brain protection measures. During brain stem function testing on two occasions she demonstrated movement in the distribution of the accessory (XI) cranial nerve despite angiographically confirmed absence of a cerebral circulation. This delayed decision making and organ donation. We report this case in order to highlight this problem which may be encountered during brain stem function testing (Patterson and McShane 1991).

Despite these clear and unambiguous recommendations, however, it is well-known that both spontaneous and reflex movements may occasionally be found in patients with brain death (Saposnik et al. 2000, 2009). The most common movements are finger jerks, although "undulating toe flexion sign, triple flexion response, Lazarus sign, pronation-extension reflex, and facial myokymia" may also be seen (Saposnik et al. 2000, 2009). For the most part, these movements have been considered to be spinal in origin, and not cortical or brainstem in origin, but regardless, they clearly constitute a challenge for the neurological determination of death, especially given that both classical and recent guidelines clearly require that the patient display no spontaneous movements whatsoever.

5.11 The Problem of Waxing and Waning of Neuronal Function

If one accepts the notion that still-living but impaired, temporarily non-working brain stem nuclei may sometimes recover to a degree as clinical conditions like cerebral perfusion pressure or arterial oxygenation vary over time, it is then possible that some brainstem nuclei will wax and wane in function during the period when brain death testing is performed.

In such a setting it is at least theoretically possible that some clinical findings supportive of brain death may vanish upon subsequent reexamination of the patient. Consequently, any loss of brainstem function identified by testing may not necessarily be permanent—one can imagine situations where a marginal brainstem nucleus later comes back into action despite not being functional during an earlier testing period.

A related concern is that a nonfunctional brainstem nucleus is not necessarily a dead brainstem nucleus; as implied above, in some cases a nonfunctional brainstem nucleus will come back into operation when conditions improve.

Admittedly, this is an issue that may be more theoretical than real, in that "recovery" from brain death is exceedingly rare. This fact notwithstanding, a case report of a very interesting example has been published by Koberda et al. (1997), who revived a patient who was "brain dead" from a thrombotic occlusion to the basilar artery (which supplies blood to the brain stem) through the timely administration of intraarterial urokinase (which unblocked the artery in question).

5.12 Limitations Inherent in All Medical Testing

If we accept the well-established notion that all medical tests have inherent limitations regarding sensitivity and specificity, and the related notion that all complex diagnostic algorithms may have inherent or unanticipated pitfalls (just as all software of sufficient complexity will manifest occasional "bugs"), then it will be understood that testing for brain death is inherently imperfect, and that type I and type II testing errors will inevitably occur from time to time, even in the setting of diagnosing brain death.

Recall that sensitivity and specificity are measures of the performance of medical tests having binary outcomes. Sensitivity is the proportion of true positives correctly identified (e.g. the percentage of people with condition X who actually have the condition). Specificity is the proportion of negatives correctly identified (e.g. the percentage of people who are correctly identified as not having condition X). These two measures are closely related to the concepts of type I and type II errors. In a type I error a patient is said to not have condition X when in fact, the patient does. In a type II error the patient is said to have condition X when, in fact, he or she does not. A perfect medical test will have 100% sensitivity (i.e. identify all people with condition X as having condition X) and 100% specificity (i.e. not falsely identify anyone as having condition X).

This view that type I and type II testing errors will inevitably occur from time to time in brain death testing is supported by both occasional reports of clinical conditions mimicking brain death as well as by the earlier discussion about possible waxing and waning of brain stem neuronal function as clinical circumstances change over time, for instance as a result of fluctuations in cerebral perfusion.

Such inherent limitations present in all medical testing imply that brain death testing has the potential for fallibility beyond mere human shortcomings in correctly carrying out brain death testing procedures.

5.13 Ambiguity in the Definition of Brain Death

In a highly-cited and highly-influential review of brain death, van Norman (1999: 275) wrote:

> Accurate criteria for death are increasingly important as it becomes more difficult for the public to distinguish between patients who are still alive from those who, through the aid of medical technology, merely look like they are alive even though they are dead. Patients and their families need to know that a clear line can be drawn between life and death, and that patients who are alive will not be unintentionally treated as though they are dead. For the public to trust the pronouncements of medical doctors as to whether a patient is dead or alive, the criteria must be unambiguous, understandable, and infallible. It is equally important to physicians that accurate, infallible criteria define death. Physicians need to know that a clear line can be drawn between life and death so that patients who are dead are not treated as though they are alive. Such criteria enable us to terminate expensive medical care to corpses. Clear criteria for death also allow us to ethically request the gift of vital organs. Clear,

infallible criteria allow us to assure families and society that one living person will not be intentionally or unintentionally killed for the sake of another.

In the following discussion I would like to make the case that, Dr. van Norman's position notwithstanding, current brain death criteria are *not* always unambiguous and are *not* necessarily infallible. In addition to the sensitivity/specificity argument presented earlier in Sect. 5.12, I offer the following additional line of argument based on imprecision inherent in clinical language.

All major protocols for the declaration of brain death contain language that may be interpreted differently by different clinical experts. One result of this fact is that there could conceivably exist situations where one expert concludes that a particular patient meets the criteria for brain death while another concludes otherwise. For instance, Canadian guidelines (Shemie 2006) indicate that the presence of "severe metabolic abnormalities, including glucose, electrolytes (including phosphate, calcium and magnesium), inborn errors of metabolism, and liver and renal dysfunction" will preclude the diagnosis of brain death. However, clinical experts may reasonably disagree as to whether a particular metabolic abnormality is truly "severe". For instance, when is hypoglycemia "severe" as opposed to merely being "moderate"?

One approach would be to look for a consensus document from the World Health Organization, the World Medical Association or other international authority. Then again, clinicians with nationalist tendencies might argue for the use of a national document as opposed to using an international reference. However, such considerations are generally moot, because such consensus documents are generally nonexistent, not yet having been developed.

One might also consult various clinical authorities for definitions. For instance, the Institute for Health Improvement suggests that hypoglycemia is severe if glucose values are less than or equal to 40 mg/dl.[7] Other authorities, however, might employ a clinical definition whereby in addition to a low glucose level, clinical findings suggestive of hypoglycemia (e.g., one or more signs or symptoms of nervousness, sweating, intense hunger, trembling, weakness or palpitations) should be present.

This simple example helps illustrate the difficulties in defining the term "severe" in a clinical context. Similar problems arise in defining the severity of other metabolic derangements such as hyperglycemia, hypernatermia, hyponatremia, hyperkalemia, hypokalemia, hypercalcemia, hypocalcemia, hypermagnesemia, hypomagnesemia etc.

This is not to suggest that such conditions cannot in principle be rigorously defined. For instance, the Canadian guidelines referenced above define hypothermia as a core temperature under 34 °C. Even then, however, the determination of core temperature will still depend to an extent on the specific measurement site employed (e.g., distal esophagus, pulmonary artery, urinary bladder, etc.).

[7] http://www.ihi.org/IHI/Topics/CriticalCare/IntensiveCare/Measures/IncidenceofHypoglycemicEpisodes.htm.

Such issues illustrate how clinical judgment is unavoidable in many such cases. And whenever clinical judgment is employed, expert clinicians will disagree from time to time. In sum, there is at least a degree of ambiguity in the language used in most brain death protocols.[8]

5.14 Further Difficulties in the Declaration of Brain Death

In a manner similar to the problem of language ambiguity, some clinicians have observed that the criteria for brain death are in some important respects somewhat arbitrary. For instance, while the original Harvard criteria for brain death requires the loss of all spinal reflexes (Ad Hoc Committee of the Harvard Medical School to Examine the Definition of Brain Death 1968) more recent criteria do not (the American Academy of Neurology 1995). Similarly, the various national guidelines are not all exactly identical, so it is likely that there on occasion one may encounter some patients who meet some existing national criteria for brain death yet do not meet others (Wijdicks 2002; Haupt and Rudolf 1999). Indeed, with the worldwide spread of transplantation surgery, numerous studies, reviews, and regional guidelines regarding brain death have been published, many with differing diagnostic criteria. As an example, the Belgian criteria for brain death requires that three physician evaluations take place, while only one is required in Poland (Haupt and Rudolf 1999). Similarly, confirmatory repetition of the testing is mandatory in France, Italy and Luxemberg, but not required in Poland or in the UK (Haupt and Rudolf 1999).

Publications from the Inter-Agency Committee on Irreversible Coma and Brain Death, the American Neurological Association, the Conference of Royal Colleges and Faculties of the United Kingdom, the President's Commission for the Study of Ethical Problems in Medicine and Behavioral Research and other sources provide a maze of criteria, cautions, exceptions, and ancillary investigations. Issues such as these have led to never ending academic discussion in the medical literature, as well as to the airing of highly controversial television programs such as one from the BBC entitled: "Transplants: Are the Donors Really Dead?" (Pallis 1980). Clearly, the details of the concept of brain death can be said to be in evolution as these and other issues continue to be discussed in various national and international forums.

Finally, it should be clear that the above issues present difficulties with respect to the notion of personhood, first introduced in Sect. 3.3. This is because being alive is a necessary (but not sufficient) condition for personhood.[9] Note that personhood, discussed in detail Chap. 3, necessitates a degree of rational activity, hence conscious-

[8]An ambiguous statement is one whose meaning is unclear, indistinct, having a double meaning, or open to more than one interpretation. It is the last sense that I argue that that there is a degree of ambiguity in the language used in most brain death protocols. This ambiguity exists in large part because some of the requirements for the declaration of brain death are specified in qualitative rather than quantitative terms.

[9]Note, however, that should non-biological self-aware organisms ever come to pass, the requirement of being alive to merit personhood will require reassessment.

ness and its associated brain activity; personhood cannot simply be equated with being a human "not being dead". There therefore exists a personhood "gray zone" between the neurologically intact human and an individual having been declared brain dead. Here I suggest that the paradigm based on the work of Hughes (Table 3.1) suffices very nicely.

5.15 The Problem of Indistinct Boundaries: Death as a Process

Yet another problem with the current approach to the declaration of brain death has been stated succinctly by Halevy and Brody (1993):

> The problem we have identified with the current consensus about brain death is not easily resolved. We believe that this difficulty is an indication of a fundamental misunderstanding in the current consensus, one that was first identified by Morison (1971) 20 years ago during the initial debate about brain death. The consensus presupposes a sharp line between life and death and tries to identify that line with one or another criterion for death. The data we have presented challenge this consensus by showing that different aspects of brain functioning cease at many different times. Thus, any sharp dichotomy between life and death based on brain functioning, although convenient and appealing, is biologically artificial.

In addition to this issue, recent decades advanced studies into the biochemistry of physiological (apoptosis) and pathological (necrotic) cell death in mammalian organisms have led to new insights into the concept of death on the macro scale, i.e., the death of the entire biological organism. As a result of these insights, some individuals have advanced the philosophical position that organismal death is not an event so much as it is a process, and that life and death are not distinct binary states, but entities with degrees and gradations. Thus, while a patient may be legally declared to be dead a few minutes after the continued cessation of cardiac activity (as is done, for example, with non-heart-beating organ donation protocols), nevertheless, these individuals argue, the cellular changes which occur in the period following cardiac arrest may be potentially reversible following the application of appropriate clinical interventions, especially under hypothermic conditions. Indeed, if all cells and organs in an organism died fully and completely immediately following the last heartbeat, any retrieved organs would be completely unsuitable for transplantation.

Based on such concepts, Bart Kosko, in his book *Fuzzy Thinking* (Kosko 1993), argues that we should think of "death in degrees" rather than death as a simple binary process. He suggests that the conceptual paradigm of "fuzzy logic", where membership in a particular class is taken as continuous rather than discrete, may be helpful in such a setting. An example may help explain this notion. A patient who has just had a surgical procedure may be asked to rate his or her pain between 0 (no pain at all) and 10 (worst pain imaginable). This pain score might be mapped into a pain "class" in some manner. For example, membership in the class "moderate pain" might take on 0% membership at a pain score of 4, and take on a 100% membership value with a pain score of 8. The class "severe pain" might be designed to start at 6 and take on 100% membership value at 10, the maximum possible pain score in the system. In the same vein, Kosko would have us talk about degrees of death, where one would have 100% death membership following cremation, but perhaps only 90% membership in the case of cryonic suspension (discussed in Chap. 6).

In a similar manner, the notion of "incomplete" brain death has now arisen in the medical literature, as might occur in the case of "the patient with a massive brain injury who meets the criteria for brain death only imperfectly, perhaps because one small patch of neurons in a brain-stem nucleus is still operating intermittently" (Doyle and Robichaud 2004). By and large, such patients are not considered to be dead and so are not considered to be suitable for organ retrieval purposes. Often, once the hopelessness of the situation is established, in the actual clinical world such patients are usually withdrawn from life support, and cardiorespiratory arrest follows inevitably some time later. Tragically, in this setting the organs are almost never used for transplantation, as they must usually be harvested prior to collapse of the circulation. Thus, the organs "go to waste" even when the patient has signed his or her organ donor card and even when the family is enthusiastically in favor of organ donation.[10]

5.16 Peter Singer and Brain Death

Readers following the debate on brain death may take an interest in the philosophical views of Peter Singer, introduced earlier. As we have seen, in his various writings he has challenged our most closely held beliefs on infanticide, euthanasia, and the moral status of animals. He also challenges the conventional wisdom on brain death.

In his book *Rethinking Life and Death* (1995), Singer notes that following the Harvard Brain Death Committee report published in 1968, most countries have adopted brain death as an acceptable criterion for declaring a person legally dead. He also notes that this event transpired with virtually no opposition despite its groundbreaking nature. What is less widely known, Singer points out, is that this "redefinition" coincided historically with the advent of organ transplantation—a mere nine months before the Harvard report came out, Dr. Christiaan Barnard performed the first successful cardiac transplant. Singer doubts that this is a coincidence.

In a manner similar to the discussions in Sects. 5.5 and 5.6, Singer likewise points out that it is simply not true that all brain function necessarily ceases with brain death—for instance, pituitary function often continues for some time after formal criteria for brain death are met.

Singer takes the position that brain dead individuals are still alive, but that organ harvesting from these individuals is none the less acceptable. His position is that rather than employ artificial, contrived, or bogus definitions of death, we should recognize that the only intellectually honest course is to admit that all lives are not equally valuable and that some lives are indeed in such a degraded and hopeless state

[10]That being said, the procedure described in Sect. 5.19 is one means to get around this problem.

that even though they are technically "alive," it is still ethically acceptable to utilize their organs for transplantation. This approach solves the problem of incomplete brain death addressed in Sect. 5.14.[11]

5.17 The Diagnosis of Brain Death Compared to a Murder Trial

The determination of brain death is in some respects analogous to a murder trial. In both cases there is a rigorous process to establish an outcome. In both cases it is theoretically possible that the incorrect outcome occurs. An incorrect outcome may occur either because the appropriate process is not followed or because of imperfections in the process. An example of the former in the case of brain death is an incorrect determination of brain death in a patient who is unresponsive because of a barbiturate overdose. In such cases an immediate determination to brain death is not permitted under brain death protocols. Similarly, in a murder trial, if a judge mistakenly disallows exculpatory evidence from being presented, an incorrect outcome may also occur.

In both the medical and legal cases, however, there is the possibility that even with rigorous adherence to the correct process an incorrect outcome may still result. This may be because the process itself has flaws, such as being based on a protocol with inherent vagueness or ambiguity. In the case of the determination of brain death, for example, expert clinicians may disagree as to what constitutes a metabolic derangement of sufficient severity that the declaration of brain death cannot be made. This is because, as indicated earlier, the language in the protocols for the declaration of brain death often use terminology that is imprecise. Thus, for example, two experts may disagree as to whether or not a serum sodium level of 127 milliequivalents per liter (the normal range is 135–145) is a sufficiently severe electrolyte abnormality that the declaration of brain death cannot be made. Unfortunately, no extant brain death protocol specifies acceptable ranges for all possible electrolyte and metabolic parameters: clinical judgment is necessarily involved.

Similarly, in a murder trial, there may be sufficient imprecision and ambiguity in the applicable legal statute under which the defendant is being tried that various judges will disagree among themselves as to whether the defendant is guilty or not.

Lawyers that specialize in legal appeals are particularly aware that mistakes occasionally occur during a trial. Similarly, mistakes occasionally occur during the declaration of brain death. Gail van Norman describes some of these mistakes in a review of the declaration of brain death published in the anesthesiology literature. For example, she describes the case of a patient who was declared to be brain dead by

[11]While the context here is that "lives not equally valuable" refers to brain dead lives in distinction to human lives generally, Singer also argues that in everyday life, some lives (not only brain dead ones) are also not equal, e.g. severely impaired *spina bifida* born infants in comparison with normally born infants.

a neurosurgeon and where the patient was brought to the operating room for organ retrieval. Towards the end of the case the heart became very tachycardic, and the anesthesiologist chose to administer the drug edrophonium to slow down the heart rate. When this happened, the patient started breathing asynchronously with the ventilator. Since any spontaneous breathing is incompatible with the declaration of brain death, the initial declaration of brain death was reviewed in detail. It was noted that the declaring neurosurgeon had observed one agonal respiratory effort during the apnea test, but he had felt that the patient was still "dead enough" to allow the declaration of brain death to be made. What makes this case especially interesting was that I was the anesthesiologist involved in this case, and this case triggered a career long interest in the problem of brain death.

5.18 Neomorts

A neomort is a brain-dead individual who still has a blood pressure and a pulse and is artificially ventilated by machine. Drug infusions are frequently used in addition to help maintain cardiovascular stability. In most cases neomorts are brought to an operating room for organ retrieval, after which ventilation is discontinued and the heart eventually stops. (Note that in most cases a formal declaration of brain death is not done on a brain damaged patient unless organ retrieval is being considered.)

However, a neomort might be of value beyond merely being a source of organs for transplantation. For example, a marginal organ (or a good organ not normally transplanted or not needed for transplantation) could conceivably still be retrieved for use in a research protocol, such as in studies of possible organ toxicity studies for new drugs. Another possible approach would to be to use an entire intact neomort as a research platform rather than to study individual organs. The importance of this possibility is reflected in the fact that drug testing using animal models have a number of serious limitations. For example, a number of drugs that tested safe on animals have proven to be harmful to humans (Vioxx and Rezulin are just two examples). Similarly, a number of drugs proven to be dangerous to animals are in fact safe for humans to use. As an everyday example, both chocolate and the popular drug Tylenol (acetaminophen, paracetamol) can be toxic to canines in doses that are completely safe for humans.[12]

Table 5.1 lists some possible research areas using neomorts. This approach might be appealing to individuals who are happy to serve as a research platform for an important question but are not enthusiastic about giving away any organs to needy recipients. In most cases, this would be for cultural or religious reasons. In addition, many animal rights activists would likely be pleased with any developments that would avoid the need for experiments involving animals. Table 5.2 lists some

[12]These facts notwithstanding, many treatments and drugs for humans and (certain) animals are the same, especially for non-human primates such as chimpanzees.

Chapter 6
Cryonic Life Extension: Scientific Possibility or Stupid Pipe Dream?

I've long thought it a pity that non-transhumanists equate trying to conquer death with a childish fear of death and a lack of wisdom. This is like saying Sir Edmund Hillary had a childish fear of mountains.

Philip Goetz.

Abstract This chapter is concerned with cryonics and suspended animation. Like the previous chapter, it contributes substantially to the philosophical debate on where the line between life and death should be drawn. The ultimate aim of cryonics is to achieve nondestructive freezing (cryopreservation) of advanced organisms like humans so that they can be safely thawed in the future, usually with a view to obtaining advanced medical treatment not currently available. In this chapter number of scientific and philosophical questions presented by cryonics are discussed, such as whether cryonics is real science or simply a waste of money, the legal and moral status of cryonically suspended individuals, and how to deal with the possibility that the thawed individual might end up with some organs working but with severe neurological impairment.

6.1 Introduction

In previous chapters we examined a variety of issues lying at the intersection of transhumanism and ethics: genetic enhancement, neuro-enhancement, human enhancement through the use of drugs, animal-human hybrids, and computer-human hybrids. Also explored were the moral, ethical and legal problems associated with brain death in both its complete and incomplete forms. It was argued that the notions of personhood and personal identity were central to the discussion.

In this chapter we explore some of the moral, ethical and legal problems associated with another process lying at the intersection of transhumanism and ethics:

D. J. Doyle, *What Does it Mean to be Human? Life, Death, Personhood and the Transhumanist Movement*, Anticipation Science 3,
https://doi.org/10.1007/978-3-319-94950-5_6

cryonic life extension, the as yet unperfected process of placing patients into a state of suspended animation using nascent technologies such as "vitrification." Once again, the notion of personhood and personal identity can be seen to be central to the discussion.

6.2 What Is Cryonics?

Important early work in the field of cryonics date back to Ettinger's seminal work *The Prospect of Immortality* (Ettinger 2005a) and his later book *Man into Superman: The Startling Potential of Human Evolution (Ettinger 2005b)*. In addition, a volume entitled *The Philosophy of Robert Ettinger (Tandy 2002)* describes the considerable impact that Ettinger had on influencing the nascent field of cryonic preservation.

The central scientific quest in cryonics is to develop, test and validate technology that would reversibly preserve brain-encoded information essential to personhood after cryonic preservation. Advocates of cryonics seek to use cryoprotectants, vitrification and extreme cold (such as immersion in liquid nitrogen) to extend the life of an individual who can no longer be sustained for long periods by conventional medical technology. This cryonic life extension is intended to be employed for decades or even centuries, until safe reversal of the preservation process as well as treatment for the underlying medical condition becomes feasible as a result of new developments in medicine and technology.

Such an undertaking is predicated on three important principles, only one of which is well-established at this time. The first principle of cryonics is that life can be stopped or suspended as long as appropriate attention is directed at preserving the basic biological structures upon which life is based. Although this principle is well-established for a number of animal species as well as for human embryos (Pegg 2007), much needs to be done to extend the principle to include full human beings. That being said, this is a field of active research in medical science and it is reasonable to expect that interesting developments will continue to be unveiled by researchers in these fields.

A second principle is that the use of cryoprotectants and vitrification will be useful in extending the first principle to larger organisms such as adult human beings. Progress on this second principle is now occurring at the level of entire organs. One particularly important scientist in this field is Gregory M. Fahy, who has achieved successful vitrification and autotransplantation of a rabbit kidney using his (proprietary) M22 vitrification solution (Fahy et al. 2009; Fahy and Wowk 2015).

The third and most controversial principle is that post-thawing damage to individual cells (especially neurons) might be repaired one molecule at a time using the emerging sciences of nanotechnology and nanomedicine. However, to the extent that cryonic preservation might eventually be achieved without damage to cellular structures, or may be achieved through naturally-occurring biological repair mechanisms, the advanced molecular repair methods referred to above may not be necessary.

Additionally, it should be emphasized that—at least in principle—damage to individual cells may not always result in loss of the information needed to reconstruct a viable post-suspension organism. Should advanced methods become available to reconstruct the entire human organism from information reliably preserved and subsequently exploited despite the presence of cellular damage following cryonic suspension, this problem might be expected to vanish. One scenario here would be to clone an entire new body but with the special proviso that the information/identity/memories of the original organism be transferred to the new organism. In such cases the patient would emerge rejuvenated and "better than well" and without the lifespan limitations of the original cryopreserved tissue. Yet another possibility, should mind up-loading and down-loading ever become feasible, is that an individual become recovered in virtual reality.[1] This possibility is discussed in Sect. 6.20.

6.3 Is Cryonics Possible in Principle?

Although so far alien to man, the ability to endure winter freezing has developed in several species of frogs and turtles as well as in a number of species of insects and microorganisms. This suggests that with appropriate scientific advances reversible freezing of larger organisms might be possible in principle.

The evidence that reversible freezing of humans may eventually be possible comes from several sources. First, there is support from nature. For instance, the wood frog *Rana sylvatica* utilizes naturally-occurring cryoprotectants in order to survive low temperatures during winter months. This natural process utilizes glucose, derived from hepatic glycogen, as well as urea, as cryoprotectants (Costanzo and Lee 2005; Costanzo and Lee 2008; Muir et al. 2007; Storey et al. 1996). As a result, these frogs can "endure freezing for at least 2 weeks with no breathing, no heart beat or blood circulation, and with up to 65% of their total body water as ice" (Storey et al. 1996). The spring peeper (*Pseudacris crucifer*) and the gray tree frog (*Hyla versicolor*) are other species that can withstand freezing temperatures during winter (McNally et al. 2003).

In recent years, the application of gene screening technology to animals that tolerate freezing has allowed scientists to identify proteins that contribute to freeze

[1] I am indebted to High Hixon, an Alcor Research Fellow, for identifying this intriguing possibility via private correspondence. Mind uploading (whole brain emulation) is the hypothetical process of transferring a conscious mind from a biological brain to a non-biological substrate (e.g., silicon computer) via as yet uninvented technical innovations (scanning, mapping, simulation). The new substrate would run a simulation so faithful to the original that it would behave indistinguishably from the native brain. For additional discussion on this fascinating issue the interested reader is invited to read an intriguing paper from Bostrom's Future of Humanity Institute: http://www.fhi.ox.ac.uk/Reports/2008-3.pdf. One interesting issue is whether or not the Heisenberg uncertainty principle in physics—which precludes one from simultaneously knowing the exact position and velocity of any particle (such as an atom)—makes this possibility impossible in principle. This issue arises because, accepting physicalism for discussion purposes, the mind is not just a set of particles with positions w, x, y, and z, but a set of particles with both position and velocity.

tolerance in animals (Storey 2004; Storey et al. 1996; McNally et al. 2003). These proteins are in turn coded for by genes (DNA sequences) such as *fr10, li16,* and *fr47* (McNally et al. 2003). As our scientific understanding of these processes improves over time, it is interesting to speculate as to whether it might be possible to use genetic engineering means to add these genes to the human genome (if they do not already exist), or to activate them should they be present in the human genome but not expressed.

The idea would be that should a patient be diagnosed with an untreatable terminal illness, genetic engineering interventions aimed at inducing natural cryoprotection might be expected to be more effective than the methods currently in use. Another possibility would be to create new babies with the required cryoprotectant genes as a risk mitigation measure against dying from accidental hypothermia in cold climate regions like northern Canada.

Although this all might seem quite outlandish, consider that as of this writing (2018) the structure of DNA has been known only six decades. Consider also that it took only seven decades for humankind to progress from the first powered flight by the Wright brothers (1903) to the first manned moon landing in 1969.

A second line of argument that the safe freezing of humans may eventually be possible stems from the fact that human spermatozoa, oocytes and even embryos are already routinely frozen in in vitro fertilization clinics around the world. Indeed, a great many individuals alive today were at one time cryopreserved embryos. In fact, two embryo cryopreservation methods are in common clinical use: slow freezing and vitrification (Kattera and Chen 2006; Loutradi et al. 2008; Kolibianakis et al. 2009).

6.4 Vitrification

Vitrification is the technology that advocates of cryonic suspension hold as having the most promise for eventual success. In vitrification, cellular water is largely replaced by one or more cryoprotectants. In addition, rapid cooling is carried out so that any remaining water is transformed directly from its liquid phase to a glassy, vitrified state that occurs with minimal formation of damaging ice crystals.

This process involves a number of special challenges. First, there are biological limitations to the degree to which cryoprotectant chemicals will be tolerated by cells undergoing vitrification. Second, while vitrification of cell clusters has been achieved, it is much harder to do this with an entire organ or an entire organism. Finally, the required cooling rates can be a challenge. For instance, for embryo vitrification, cooling rates between 15,000 and 30,000 °C/min are often used (García-Velasco and Pellicer 2007).[2] Despite these challenges, as noted earlier, the successful vitrification and subsequent transplantation of a rabbit kidney by Fady (Fahy et al. 2009; Fahy and Wowk 2015) suggests that eventual success is mostly a matter of effort, money and imagination.

6.5 Substrate Preservation Versus Information Preservation

In dealing with cryonics, it is helpful to distinguish between the information preserved in a person's brain and the substrate used to hold that information. The information that makes us a person includes our longitudinal memories, as well as our hopes, our dreams and countless other mental events. As we discussed in Sect. 3.3, this information is unique to us—it is what makes us persons. How exactly this information is encoded into the brain substrate is a field of active neurologic research. The reader interested in this complex question might start by looking at some of the articles published in the academic journal *Learning and Memory.*[3] For individuals who would like a brief sketch on this matter, the following commentary is offered.

One popular theory of how memory is encoded in the brain is based on a hippocampal synaptic plasticity model (Bliss and Collingridge 1993; Shapiro 2001; Dudai 2004; Conner et al. 2009). The support for this theory comes from a number of sources. First, in Alzheimer's disease the hippocampus is one of the earliest brain structures to suffer damage. That memory problems appear among the first clinical findings in Alzheimer's disease suggests that the hippocampus has an important role in memory formation.

Second, lesions of the hippocampus in humans may prevent the acquisition of new memories, as in the well-known case of HM. In 1953, a man known as HM lost substantial portions of his hippocampus, parahippocampal gyrus, amygdala and anterolateral temporal cortex in a neurosurgical procedure aimed at stopping his intractable epileptic seizures. While the surgery was successful in eradicating the seizures, HM ended up with complete anterograde amnesia, although his working memory and procedural memory remained intact.

Third, hippocampal synapses are known to have "activity-dependent synaptic plasticity" that provides neuronal level evidence of a possible role in memory. In

[2]For vitrification on the human-scale, thermal mass and heat conduction properties conspire to limit the rate of cooling, with the consequence that novel approaches to vitrification must be sought. Recognizing that there is non-covalent competitive binding of water molecules between the biomolecules (which incorporate water as an integral component of their structure), the various components of the cryoprotectant, and crystalline ice, may be the basis for future developments. Note that the goal of cryoprotection is to avoid irreversible alteration of biomolecular structures, which ice does by competitive dehydration; the affinity of the water molecules for the ice is greater than for the biomolecules. (I am indebted to High Hixon, an Alcor Research Fellow, for making this point during our private correspondence.)

[3]Information about the journal is available at http://learnmem.cshlp.org.

this arena, a particularly attractive family of molecular candidates for modulating synaptic plasticity during learning and memory processes are neurotrophins (NTs).

Finally, the importance of the hippocampus to neuroscience in general and memory research in particular is evidenced by the fact that there is a journal devoted specifically to it (called, appropriately, *Hippocampus*).[4]

A computer analogy may be helpful in explaining this distinction between the information preserved in a person's brain and the substrate used to hold that information. In order for a computer to be useful its hard drive (or equivalent memory) must be loaded with a number of kinds of information. For example, the hard drive will contain an operating system, as well as a number of applications (word processing, spreadsheets etc.), as well as a number of data files (such as the text file containing these very words). For the computer to be useful, it must contain both the hardware as well as the information that constitutes its "identity" (programs and data). If the computer is destroyed, for example by fire, mere replacement of the hardware would be insufficient to restore the computer to its previously useful state. Instead, all the information that was unique to that computer would have to be added as well.

In the world of computers, restoring this information is a relatively simple task using backup programs that are commonly available. In the case of the brain, there is as yet no way to achieve this task. In the realm of science fiction brain backups are done using methods whereby the state of the brain is measured in sufficient detail that it can be reconstructed on demand. In this sense, there is a rough analogy to teleportation, another popular science fiction theme.

The importance of this distinction between the substrate and the information encoded on the substrate is that it could turn out that cryonic brain preservation ends up preserving brain tissue (substrate) but might still not be sufficient to preserve the information the brain retains.

6.6 The Two Usual Cryonics Scenarios

In the USA, Alcor, a facility located in Scottsdale, Arizona, is one of several companies offering a cryonic suspension service. (Although other cryonics organizations exist, their services, policies, and clinical approaches are similar.[5]) Candidate patients carry an alert bracelet requesting that as soon as possible after death, a large dose of heparin (an anticoagulant) be given intravenously and the newly "dead" patient be placed on cardio-pulmonary bypass to allow continuing organ perfusion prior to freezing.

---

[4]Information about the journal is available at http://www.wiley.com/WileyCDA/WileyTitle/productCd-HIPO.html.

[5]A listing of various services is available at http://www.benbest.com/cryonics/CryoFAQ.html#_VII__.

Two options are offered: a whole-body option (cost: $200,000 and up) and a less expensive option where only the head is cryonically suspended (cost: $80,000 and up). In either case the hope is that at some future time the body (or head) will be able to be unthawed and repaired, although in the case of the head only option, the problem of finding a matching body exists unless: (1) a new body can be constructed (perhaps via a variant of cloning or by advanced tissue engineering methods), or if (2) the newly reanimated head is configured to exist as an isolated perfused preparation (vide infra), or (3) the head's information contents are transferred to an entirely new biological entity (again, perhaps constructed via a variant of cloning), or (4) somehow a new organism is made to exist entirely in silico.

6.7 Living as an Isolated Head

In the situation where only the head would be preserved, one can imagine three possible outcomes: (1) failure to successfully reanimate the individual in question (certainly a possibility, at least in the early experimental phases of any cryonics research), (2) existence as a conscious, reanimated head attached to a new body (a situation with numerous challenges beyond mere successful thawing of the head), and (3) existence as a conscious, reanimated head existing independently from a body via an artificial cardio-pulmonary-renal-endocrine support system.[6]

About 50 years ago, White et al. were able to isolate a series of five monkey brains by surgically removing all anatomical structures surrounding each brain except a small basal plate of bone and the central portion of the skull (White et al. 1963, 1964, 1965, 1996). These brains were completely isolated from the donor monkey's body neurogenically and vascularly, and were perfused in vitro for 30–180 min via extracorporeal circulation using an immunologically compatible second monkey. That the isolated perfused brain remained alive and working in this setting was demonstrated by showing persistent electro-encephalographic activity at the cortex as well as by showing that the brain was appropriately extracting oxygen from its blood supply (by obtaining expected differences in oxygen content between the blood entering the brain and the blood leaving it).

A year later White's team successfully repeated the experiments using a mechanical extracorporeal system (instead of using the body of another monkey as a life-support system for the isolated monkey brain). In addition, White's team was able to transplant the head of one monkey onto the body of another. The procedure was a success to the extent that the transplanted head was able to smell, taste, hear, and see,

[6] Although this last possibility—that of a conscious human existence somewhat like the "brain in a vat" scenario so often discussed in introductory philosophy courses - seems to be almost ludicrous, the situation may be technically possible, at least for moderate periods of time. Regardless, it should be emphasized that the usual "brain in a vat" scenario is a philosophical thought experiment often employed as an argument for philosophical skepticism and solipsism, and has nothing to do with cryonics.

even though the animal was quadriplegic (since the spinal cord was not connected). The animal even occasionally tried to bite some of the staff.

Of interest, White's team was not the first to show that conscious existence as a severed head was possible. Many years earlier, in 1928, Bryukhonenko and Tchetchuline at the Institute of Experimental Physiology and Therapy in the USSR showed that life could be maintained in the isolated head of a dog for a moderate period of time by connecting the carotid arteries and jugular veins to a system for extracorporeal perfusion and oxygenation that used the lungs from a second dog as part of the setup (Brukhonenko and Tchetchuline 1929a, b; Adamenko 1969). Bryukhonenko went on to pioneer a number of important clinical developments that lead to the first Soviet open-heart operation in 1957. For these achievements, Bryukhonenko was posthumously awarded the Lenin Prize.

Evidence that the isolated perfused dog head was alive and responsive to stimuli comes from the following New York Times report (Kaempffert 1943):

> When the eye was touched it twitched. After twenty minutes there were more signs of life. The eyes were open by that time and looked alive. The head responded to a whole series of stimuli. Eyelids blinked when hairs on the brow were plucked. Particularly noticeable was the response when the mucous membrane of the nose was irritated. In fact, it was often necessary to hold the head on the plate by force. The muzzle was opened and the teeth were bared in a snarl. When quinine was placed on the tongue there was every sign of repugnance. Pieces of sausage were swallowed and ejected through the top of the alimentary canal. In short, the head behaved just as if it were attached to the body. And in this condition it remained for about three and a half hours.

A 1940 public domain film "Experiments in the Revival of Organisms" demonstrating the experiment is available at the Prelinger Archive (http://www.archive.org/details/prelinger). The film may also be viewed at http://www.youtube.com/watch?v=ap1co5ZZHYE.

6.8 Head Transplantation

In 2013 Dr. Sergio Canavero, director of the Turin Advanced Neuromodulation Group in Turin, Italy, announced plans to launch a human head transplant program (Canavero 2015; Canavero and Ren 2016; Li et al. 2017). Given White's pioneering work, many of the surgical details related to head transplantation have been worked out. In addition, White's work has recently been replicated in China as a means to further move ahead with head transplantation.[7]

Still, Canavero and co-workers face two significant obstacles that will have to be overcome: [1] fusing the donor and recipient spinal cord into a neurologically integrated unit, and [2] preventing rejection of the graft (an interesting question arises here—is the graft the body or is it the head?).

[7] http://www.iflscience.com/health-and-medicine/surgeons-claims-head-transplant-monkey-has-been-successfully-carried-out/.

Experience in composite graft (e.g., face) transplantation suggests that the immunological problem, while undoubtably formidable, can likely be overcome, so getting the recipient and donor spinal cords to fuse meaningfully will be the central obstacle to deal with. However, given that polyethylene glycol has been shown to prompt the growth of spinal cord nerves in animals (Borgens and Shi 2000; Ren et al. 2017; Kim et al. 2017), this impediment may eventually be manageable as well.

Some proposed technical refinements that extend White's pioneering work have been offered by Canavero, the lead scientist/evangelist in the initiative (Canavero and Ren 2016). First, the recipient's head and the donor body will undergo therapeutic hypothermia to extend the time that the tissues are exposed to any necessary hypoxic/ischemic insults related to the procedure. Cleanly severing both spinal cords so as to avoid ragged ends is another key step. Following approximation of the donor and recipient ends of the spinal cords, implanted electrodes would be used to provide therapeutic electrical stimulation to the spinal cord (Mondello et al. 2014). Also, should the application of polyethylene glycol fail to do the job, the use of stem cells (Tukmachev et al. 2015; Schroeder et al. 2015; Zhao et al. 2015; Li et al. 2014; Ren et al. 2017) or olfactory ensheathing cells (Tabakow et al. 2014) are two additional therapeutic approaches that the team is considering.

Of interest, Canavero's team is not the only group working on this formidable challenge. A Chinese team (Ren et al. 2014; Ren and Laugel 2013) continue to work on allo-head and body reconstruction in a murine (mouse) model. Their novel approach involves "retaining the donor brain stem and transplanting the recipient head", an approach that has the advantage of preserving respiratory and circulatory function. Recently, Canavero has been formally collaborating with Ren and other Chinese scientists.

Undoubtable, a number of ethical issues are involved in going forward in early human trials. For example, what should be the role of animal trials before moving forward to humans? Consider, for instance, item 3 of the Nuremberg Code, which states that prior animal experimentation be used before moving to human trials: "The experiment should be so designed and based on the results of animal experimentation and a knowledge of the natural history of the disease or other problem under study that the anticipated results will justify the performance of the experiment."

A completely different ethical issue has been raised by Corlett (2001), although not in this specific context. He makes the argument that we have a duty to die inexpensively rather than squander money on a therapy with very limited utility (as would very possibly be the case with head transplantation). Basically, in a setting of limited health care resources, argues Corlett, we have a duty to spend the money only where the impact will have the greatest value.

Let's end this section with a specific hypothetical scenario. A famous quadriplegic scientist has developed terminal body failure and, having only months to live, requests that Dr. Canavero attach his head to the body of an available brain-dead man whose family is agreeable to donating more than just the customary organs. The scientist knows the risks as fully as anyone might, and while he is hopeful for the return of full spinal cord function, quadriplegia is an outcome he can (and does) live with.

The family is honored by the possibility that their son's body will be so wonderfully recycled. You are on the hospital's ethics board. How do you vote?

6.9 An Interesting Hypothetical Scenario

Let us now consider two thought experiments, presented in this and the following sections. As with most philosophical thought experiments, these are offered to the reader as a method to explore the consequences of particular philosophical positions.

It is 2045. Generalissimo Sanchez rules the nation of Sanchiva with an iron fist. Unfortunately for him (but perhaps not for his subjects) he is diagnosed with metastatic melanoma with an exceedingly poor prognosis. He now has only a few months to live, despite having access to the best possible medical care and a massive personal fortune. To everyone's surprise he decides to step down and put his eldest son in power. Even more surprising, he decides to be cryonically suspended before his anticipated death, with a view to be reanimated at some future time when effective treatment becomes available. To this end, he hires a team from the Mayo Clinic and from Alcor to induce general anesthesia, establish anticoagulation and cardiopulmonary bypass, and then freeze his body under general anesthesia according to the Alcor protocol, which now has advanced to the point that their technology to prevent cell damaging ice crystal formation has become very reliable (at least in animal studies). Following this, the Generalissimo has his frozen body placed in a special mausoleum featuring triply redundant cooling systems and advanced electronic monitoring, as well as a well-guarded viewing station for the convenience of the populace.

This bizarre and macabre scenario raises a number of interesting questions: Is the Generalissimo Sanchez now dead or alive? What moral standing does the Generalissimo now have? How would one gauge the success of any reanimation efforts?

6.10 A Second Hypothetical Scenario

Field Marshal Perrera has adeptly but firmly ruled the oil-rich European island nation of Novacia since he overthrew the democratically elected but corrupt government over a decade ago. He now wishes to institute a uniquely bizarre form of capital punishment. To avoid criticisms that his shady judicial system might kill innocent citizens via capital punishment, he sentences people found guilty of capital crimes to indefinite cryonic suspension. That way, should exculpatory evidence become available, clearing the accused and resulting in a more favorable judicial outcome, there is at least a theoretical chance of a post-thaw recovery. On the grounds that a better outcome would be expected if the freezing is done before death has actually occurred, a protocol identical to that described in the earlier scenario is used.

As in the previous scenario, some interesting questions are raised. What are the legal and moral standings of such cryonically suspended individuals? Based on the argument that the suspended individuals are not really dead, would such an arrangement still be acceptable if Novacia sought admission to the European Union (which forbids capital punishment)?

6.11 Cryonics as "Borderlands Science"

The information and scenarios presented so far raises a number of intriguing scientific, philosophical, social and ethical questions. Let us explore some of these questions, starting with the question of whether cryonics is real science.

Many professional cryobiologists do not regard cryonics as worthy of serious study, particularly when the possibility of nanobots carrying out post-thaw cellular repairs is raised. For instance, since cells contain thousands upon thousands of components that may be damaged, some authorities argue that developing the required nanobot for each conceivable type of cellular damage may be impossible, or at least highly impractical. As another example of the skepticism raised by scientists, consider the words of Shermer (2001) writing in *Scientific American*:

> Cryonicists believe that people can be frozen immediately after death and reanimated later when the cure for what ailed them is found. To see the flaw in this system, thaw out a can of frozen strawberries. During freezing, the water within each cell expands, crystallizes, and ruptures the cell membranes. When defrosted, all the intracellular goo oozes out, turning your strawberries into runny mush. This is your brain on cryonics.[8]

Shermer also writes:

> This is what I call "borderlands science," because it dwells in that fuzzy region of claims that have yet to pass any tests but have some basis, however remote, in reality. It is not impossible for cryonics to succeed; it is just exceptionally unlikely. The rub in exploring the borderlands is finding that balance between being open-minded enough to accept radical new ideas but not so open-minded that your brains fall out. My credulity module is glad that some scientists are devoting themselves to the problem of mortality. My skepticism module, however, recognizes that transhumanistic-extropian cryonics is uncomfortably close to religion. I worry, as Matthew Arnold did in his 1852 poem "Hymn of Empedocles," that we will "feign a bliss/Of doubtful future date,/And while we dream on this/Lose all our present state,/And relegate to worlds yet distant our repose.

Shermer is arguably unfair, however, in some of his criticisms of cryonics, especially in his strawberry analogy. He fails to mention, for instance, that it is the goal

[8]In cryonics circles this is known as the "strawberry man" argument. Needless to say, this is NOT what happens with successful cryonic vitrification. Shermer's failure to mention vitrification in his article, despite it being the center-piece of cryonics, and the refusal of the editor of Scientific American to publish a letter to this effect, is an example of publication bias in science. For additional commentary on this and related issues, the interested reader is directed to http://www.alcor.org/press2001SciAm.html and http://www.icsu.org/publications/cfrs-statements/bias-in-science-publishing.

of vitrification science to avoid irreversable denaturation of cell biomolecules at low temperatures, and that this process does occur in nature in some organisms (as discussed earlier in this chapter).

6.12 Cryonic Reincarnation?

Another issue is that it is conceivable that a thawed individual might end their lengthy cryonic slumber with a substantial degree of information loss, so that the longitudinal memories that define us as persons would be substantially gone. The thawed individual would in a way have become a "Tabula Rasa", ready for "reprogramming" with new experiences and memories but unaware of his or her past. This is a possibility I call "cryonic reincarnation." To the extent that the long-term goal of cryonics is to develop technology to reversibly preserve brain-encoded information essential to personhood, it should be clear that cryonic reincarnation would actually be a form of failure. After all, the goal of cryonics is to preserve both the substrate and the information held by the substrate—mere preservation of the substrate alone, while a marvelous biotechnical achievement is still a failure of cryonics. Also, the simpler objective of mere substrate preservation could likely be achieved by human cloning, a process much simpler than cryonic preservation.

6.13 Some Philosophical Considerations

Smith (1983) has argued that both cryonics and genetic engineering, in replacing natural processes with technological processes, "disturb the delicate balance of the triad of life which each individual experiences–faith, health, and justice." As a consequence, Smith argues, these new biological technologies raise a number of difficult legal, social and philosophical issues, such as how should the legal system deal with faith and religion issues surrounding the cryonics patient and how should it respond to the cryonics physician "who views death as a disease which is curable?" One problem with Smith's analysis is that from the dawn of time human beings have been replacing natural processes with technological processes. Would Smith have us live like Stone Age creatures? The limiting cases are "in a state of nature", which humans departed from many thousands of years ago, and—at the other extreme—"anything allowed by physics".

One particularly interesting issue concerns the legal and moral status of cryonically suspended individuals. It is completely unwarranted to argue that cryonics patients *will* certainly be resuscitated in the future. On the other hand, it would be wrong to unconditionally rule out the possibility of eventual success. In this sense, cryonically suspended individuals lie somewhere on a continuum between being unequivocally alive and being irreversibly dead. However, neither our legal system nor our conventional approaches to medical ethics are currently well-equipped to deal with uncertainties of this kind.

Another unresolved issue concerns the legal and moral status of any thawed individuals who end up technically alive but with severe neurological damage, such ending up in a chronic vegetative state (vide infra).[9]

6.14 Some Bioethical Considerations

As discussed earlier, central to cryonics is that notion that life and death are not always simple binary events. Similarly, some authorities have argued that the classic notion of basing the definition of death on cessation of the heartbeat is philosophically problematic. This is because, as discussed earlier in Chap. 5, death is not an event that happens instantaneously when the heart stops, but rather is a complex process involving a series of biochemical events at the cellular level that takes time to complete once tissue perfusion stops as a result of the cessation of cardiac contractions.

In fact, in cases where the heart can be restarted, the clinical result may sometimes still be excellent, without any obvious neurological impairment or damage to other organs. For instance, cold water drowning victims have been resuscitated without clinical sequellae after periods of submersion and cardiac arrest in excess of 30 min. Some critics have argued that the concept of death based on classical criteria such as cardiac asystole or resuscitation failure is in fact a mere social construction lacking a rigorous scientific foundation. Consider the following commentary from Whetstine et al. (2005): "…few if any patients pronounced dead by today's physicians are in fact truly dead by any scientifically rigorous criteria." It is with this approach to death that cryonicists work, hoping to cryopreserve their patients before sufficient brain damage has occurred that absolutely irreversible death has set in.

In this regard, Hershenov (2003) has emphasized that most definitions of death—whether cardiopulmonary or on neurological grounds—implicitly require irreversibility. Thus, mere cessation of cardiac or neurological function should be insufficient grounds to declare death. However, current clinical rules for the declaration of death do not actually require that irreversibility be rigorously established. For instance, in the case of organ donation after cardiac death, it might well be possible to restart the heart after the required period of asystole for the declaration of death has passed. And even if the heart might not be restartable, one could emergently institute cardiopulmonary bypass and install one of the fully implantable artificial hearts now commercially available. (In my opinion, the AbioCor artificial heart would show special promise in such a scenario).

---

[9] Arguably, though, before thawing is attempted, the brain should be repairable to the extent that basic functioning would likely be fully restored, even if pre-preservation memories are not. And such memories in turn might be reconstructed to an extant from records, especially if the patient made a special effort to record past experiences, etc. (I am indebted to Mike Perry, Ph.D., an Alcor employee, for making this observation to me in private correspondence.)

Hershenov also emphasizes that the thorny issue of the definition of death likewise arises in the context of cryonic suspension: "in the future, millions of cryogenically frozen human beings could spend centuries in a non-dead state because of the future technological breakthroughs; or large numbers of 'frozen' people are dead for eons but coroners are not able to declare them so because they are unaware of what biological conditions science will never be able to reverse."

It would thus appear that true irreversibility of death is incompatible with successful cryonic revival, but fortunately for enthusiasts of cryonics, true irreversibility is not always present at the time that death is declared in most clinical situations. Accordingly, cryopreservation with the possibility of future resuscitation requires a "window of opportunity" between legal death and irreversible death.

6.15 Some Negative Bioethical Considerations

One aspect to the debate hinted at above but not emphasized in the cryonics literature is the very real possibility that any initial "successes" with cryonic reanimation may be accompanied by severe neurological injury, such that the revived individual would remain comatose or in a minimally conscious state or exhibit other profound impairments that would render life too degraded to be worth living. The result would be that the revived individual would not get the outcome that was hoped for, with the additional complication of now having a patient who requires complex medical care with very little likelihood of improvement.

Possible counterarguments to this concern are twofold: first, the expectation is that a "last in, first out" process will take place where the oldest cases will be the most difficult and the last to be attempted. Secondly, if technology able to resuscitate cryopreserved people comes to be developed, then it is hoped that other companion technologies will also become available so that the worst likely deficit will be some degree of amnesia.

Other potential objections to cryonics exist. For instance, Shaw (2009) has summarized some of the practical and ethical objections to cryonics: it "would change the very concept of death"; "it is 'against nature'"; "no friends or family of the 'freezee' will be left alive" following thawing; the science of cryonics "might not advance enough to ever permit revival"; "reanimation might not take place due to socio-political or catastrophic reasons"; and "cryonics could lead to premature euthanasia in order to maximize chances of success."

After reviewing these and other arguments, Shaw offers a variation of Pascal's Wager as an argument in favor of utilizing cryonic suspension. (Recall Pascal's Wager: even though the existence of God cannot be established through pure recourse to reason, one should still live life as though God exists, because in doing so one has eternal life to gain, and little to lose.)

However, I disagree with Shaw in this matter. My clinical opinion is that the *likely* outcomes following thawing, at least in the early years, will be predominantly negative, consisting variously of a recovered individual with no memories, a recovered individual with severe neurological impairment or a thawed individual with some organs working but no brain function at all. Pascal's Wager is worthwhile only if there is little to lose if the wager is lost. In my opinion, these likely negative outcomes hardly constitute "little to lose."[10]

6.16 Burden on Society?

Dr. John Baust has made the comment that an "individual who freezes himself or herself to come back in the future makes the assumption he will be a contributor to that society and that they would want him".[11] Such a comment carries with it both interesting practical and ethical perspectives. The practical aspect here is that the recovered individual might require complex, lengthy and expensive medical care to fix the problems that led to them undergoing cryonic suspension in the first place unless means such as growing new bodies via tissue engineering become perfected. Patients who were frozen as severed heads would have the special challenge of finding (or constructing) a body to which their head would be attached. Also, funding to deal with all of these issues would likely not be in place, resulting in a burden on society.

The ethical issues involved in Baust's comment run deep. Many individuals, especially those who regard human life as being sacred, would take issue with the implied notion that human beings have value to society only to the extent that they are wanted (for example by family members) or are able to make social contributions. On the other hand, as noted in Sect. 6.8, Corlett (2001) argues that in health care contexts "there is a moral duty to die inexpensively." Certainly, cryonic suspension would be viewed by a great many of its detractors as a particularly expensive way to die.

6.17 A Waste of Money?

Dr. Kenneth Goodman, a bioethicist writing in the Miami Herald, has commented "If you have enough money (for cryonics), then you have enough money to help somebody in need today." [12] In a similar vein, Jean Medawar is said to have commented "Money invested to preserve human life in the deep freeze is money wasted,

[10]I should point out that many in the cryonics community don't think these are "likely outcomes." They argue that should cryonic resuscitation become possible, contemporaneous technological developments will also make it possible that an individual will awaken without neurological deficits.

[11]Quoted at http://www.alcor.org/notablequotes.html.

[12]Quoted at http://www.alcor.org/notablequotes.html.

the sums involved being large enough to fulfill a punitive function as a self-imposed fine for gullibility and vanity."[13]

Both these comments raise interesting ethical issues. The first comment about not spending money on membership in a "cryonics club" so that others might benefit from that same money applies equally well to a number of more expensive expenditures that most ethicists do not criticize. For instance, as of this writing, the annual cost of membership with Alcor for adults in the U.S. is $800. Additional adult family members pay $490, minors, $156. This is in addition to the whole-body preservation cost of $200,000, which is often achieved from a life insurance policy payout. While these are not small sums, many Americans maintain (for example) luxury estates or second vacation homes at much higher costs than that of a cryopreservation plan without being seen as being unethical for not having given this money to help others.

On the other hand, Peter Singer has argued exactly this—that we in the West are morally obligated to live a relatively modest life, living below our means so that the resulting discretionary money can be used to substantially improve the lives of others. In his book *The Life You Can Save: Acting Now to End World Poverty* (Singer 2009), Singer suggests that we are not required to choose between asceticism and self-indulgence, but that a middle solution based on unpretentious living can free up money to allow us to donate to charities where a mere few hundred dollars can save a life.

Singer provides the following example to make his point. Imagine that you are walking in the park and you find a little girl drowning in a water fountain. Do you help? Almost everyone would agree that would be the right thing to do. Similarly, argues Singer, the same reasoning applies to helping solve global poverty. The only difference is that in the former case the person being helped is located directly in front of you while in the other case the person being helped is far away.

A related ethical issue arises with the suggestion implied by the second comment (by Medawar) that paying for cryonics services is a form of "fool's tax" or a "tax on hope". But arguably similar issues arise with other experimental endeavors with very uncertain outcomes. Would one label Louis Washkansky, the recipient of the first heart transplant (Cape Town 1967), as being "gullible and vain" because he took a great gamble but lived only 18 days after the transplant? Similarly, does the use of newly developed genetically engineered drugs that cost thousands of dollars per dose yet are not fully guaranteed to be helpful constitute a waste of money? (Burrows et al. 2002).

6.18 Some Legal Considerations

The legal issues that arise in cryonics will vary from location to location, and usually depend to a large extent on applicable definitions of death and protocols regarding

[13]http://people.ucalgary.ca/~kmuldrew/cryo_course/cryo_chap15_1.html#02. Note also that many have said the same for donating to religious institutions.

interment. In the USA a particularly interesting legal case pertaining to cryonics was the case of Donaldson versus Van de Kamp (Pommer 1993). Donaldson, suffering from an inoperable brain tumor sought the legal right to obtain antemortem cryonic suspension of his body, knowing that this was technically preferable to a postmortem suspension. However, both the trial court and the appellate court ruled against Donaldson, arguing that the cryonic process in this case would be a form of physician-assisted suicide, and that Donaldson did not have a constitutional right to such a form of death. Specifically, the court determined that "the state's interest in preserving life, preventing suicide, protecting innocent third parties, and maintaining the ethical integrity of the medical profession" trumped Donaldson's interest in cryonic suspension until treatment for his brain tumor was available (California. Court of Appeal, Second District, Division 6 1992).

This case suggests, however, that should physician-assisted suicide ever become legal in the USA, antemortem cryonic suspension, with its greater likelihood of favorable outcomes compared to postmortem cryonic suspension (at least theoretically), could be legally practiced.

6.19 Religious Perspectives

Like a belief in an afterlife found in many religions, participants in cryonics take a leap of faith that future resuscitation with a worthwhile outcome will be possible. In truth, however, the future repair technologies assumed by cryonics enthusiasts remain at best hypothetical and are certainly unproven, in a manner similar to a number of religious beliefs.[14]

Likewise related to religion, for many thinkers a fundamental issue is whether cryonic suspension should be regarded as a form of interment or as a form of medical intervention. Some religious thinkers who define death in terms of the soul leaving the body argue against cryonics because "only God can resurrect the dead". However, a potential counterargument that can be offered to believers in such a setting is that the soul leaves the body only with the onset of *irreversible* death, and, as with the above discussion concerning organ donation after cardiac death, a "window of opportunity" between ordinary clinical death and irreversible death once again can be seen to exist, with the soul fully leaving the body only with the onset of irreversible death.

Chapter 8 discusses some other religious issues related to transhumanism.

[14]That being said, cryonicists' expectations fall within the realm of physical law, while most religions place faith in the supernatural.

6.20 Whole Brain Emulation and Mind Uploading

"Whole brain emulation" (also known as "mind uploading", "mind copying" or "mind transfer") is a term used by transhumanists to refer to the emulation of a brain on a (presumably digital) computer.[15] In essence, it refers to the hypothetical process of transferring a conscious mind from a biological brain to a non-biological substrate (e.g., silicon digital computer) via technical developments (scanning, mapping, simulation) that are as yet uninvented. The idea is that the new substrate would run a simulation so faithful to the original that it would behave indistinguishably from the native brain, complete with the emergence of consciousness along with one's memories and experiences.

Two possible hypothetical methods are envisioned. In the first hypothetical method, mind uploading would be achieved by high-resolution 3D scanning, mapping, modeling and characterization of all relevant features of an individual's biological brain, and then running the derived brain model in a high-powered host computer system.

In the second hypothetical method neurons in the biological brain would be gradually replaced by synthetic versions until the individual's brain would be entirely synthetic.

In either case, some transhumanists view the possibility of whole brain emulation as an important putative life extension technology, either as an alternative to cryonic preservation or as a technology that would serve as a supplement to cryonic preservation in cases where unrecoverable brain damage is encountered after thawing.

Whole brain emulation has also been proposed as one possible approach to advance progress in Artificial Intelligence, the idea being that advances in whole brain emulation will allow for developments in Artificial Intelligence not possible by ordinary means, such as the creation of conscious, sentient artificial beings.

It should be emphasized, however, that many transhumanists are skeptical of these possibilities. Some argue that whole brain emulation is not possible in principle, believing that artificial intelligence equivalent to human intelligence—so-called strong AI—is not possible, i.e. not computable.

Other thinkers point out that the operation of any biological brain depends on time-varying electrical (action potential) and biochemical (neurotransmitter) signal exchange between dendrites, axons and other neuronal structures, so that capturing the brain state in a single "frozen" state will likely prove to be insufficient to fully capture the information needed for whole brain emulation.

Still others believe that whole brain emulation may be contrary the laws of physics—specifically, the Heisenberg Uncertainty Principle. (The Heisenberg Uncertainty Principle precludes one from simultaneously knowing the exact position and

---

[15]Highly readable popular accounts are available at https://www.popsci.com/article/science/neuroscientist-who-wants-upload-humanity-computer as well as in book form: Keith Wiley, Taxonomy and Metaphysics of Mind Uploading, Alautun Press, 2014, available at https://www.amazon.com/dp/B00NJZHGM8.

velocity of any particle (such as an electron) at any chosen point in time. Consequently, accepting mind physicalism for discussion purposes, the mind is not just a set of particles with positions w, x, y, and z, but a set of particles with both position and velocity.) Thus the idea that future technology might allow one to scan the brain in precise detail such that the characteristics of every neuron, synapse, and dendrite and the position and velocity of every neurotransmitter molecule and electrical impulse in the human brain would be known exactly would likely be contrary to the laws of physics.

Finally, some thinkers believe that even if whole brain emulation is possible in principle, the required resources for whole brain emulation will never become obtainable within the anticipated life of the universe.

Whole brain emulation is associated with some philosophical issues not unlike those discussed earlier in this chapter in the context of cryonic suspension. Whole brain emulation is predicated on the notion of mind physicalism, that the human mind can be adequately represented by axons, dendrites, synapses etc. This mind model is thus potentially in opposition to the dualism model offered by Descartes and other philosophers of mind who would have us possess an immaterial soul in addition to a material body.

Another issue, discussed earlier in Chap. 3, concerns the rights and responsibilities of newly created conscious, sentient artificial beings produced via brain emulation technologies. Related to this is the concern that partially successful whole brain emulation might conceivably produce a "philosophical zombie" that is excellent at symbol manipulation (as in Searle's Chinese Room argument) but is not in possession of sentience or consciousness.

In closing this discussion, it should be noted that such philosophical issues are surprisingly old. In 1775 Thomas Reid wrote in a letter to Lord Kames[16]:

> I would be glad to know your Lordship's opinion whether when my brain has lost its original structure, and when some hundred years after the same materials are fabricated so curiously as to become an intelligent being, whether, I say that being will be me; or, if, two or three such beings should be formed out of my brain; whether they will all be me, and consequently one and the same intelligent being.

6.21 Conclusion

In conclusion, the possibility of human cryonic suspension, while admittedly remote and far-fetched to many, raises a number of fascinating philosophical and ethical questions (Smith 1983; Tandy 2002). Special concerns discussed in this chapter included the matters of moral standing, the nature of personhood and what it means to be a living human being. It was emphasized that central to the possibility of successful human cryonic suspension is the idea of a "window of opportunity" between legal death and irreversible death—a clearly enticing philosophical notion. Of course,

[16]http://www.benbest.com/philo/doubles.html.

even better from a purely technical point-of-view would be antemortem cryonic suspension.

Chapter 7
Defending Attacks Against Transhumanism

All truth passes through three stages. First, it is ridiculed. Second, it is violently opposed. Third, it is accepted as being self-evident.

Arthur Schopenhauer, German philosopher (1788–1860) (As is the case with some earlier quotations used at the beginning of previous chapters, this quotation is offered as "food for thought" and not as a philosophical stance defended in the chapter.).

Abstract This chapter is one that is particularly rich in philosophical arguments and has as its main focus the critical evaluation of the arguments against transhumanism that have been raised by a number of prominent philosophers and other individuals. Based on the critiques of transhumanism offered by critics such as Leon Kass, Francis Fukuyama, and Michael Sandel, the chapter critically assesses five commonly raised arguments why human enhancement by artificial means should not be undertaken: fairness, human dignity issues, "slippery slope" arguments, the "repugnance" argument, and safety issues. The weakness of "slippery slope" arguments and "repugnance" arguments are given special attention. In the end it is concluded that safety issues are the most realistic concerns that the pursuit of transhumanism raises.

7.1 Introduction

Transhumanism remains a subject of considerable controversy among thinkers. As discussed earlier, the principal aim of transhumanism is to develop and make widely available ethical human enhancement technologies that will improve the human condition. Thinkers favoring the wise use of human enhancement technologies are individuals such as James Hughes, author of *Citizen Cyborg*, Nick Bostrom and Julian Savulescu of Oxford University, John Harris of the University of Manchester, as well as many other lesser known members of the transhumanist community. Such individuals favor the voluntary, rational deployment of human enhancement technologies

D. J. Doyle, *What Does it Mean to be Human? Life, Death, Personhood and the Transhumanist Movement*, Anticipation Science 3,
https://doi.org/10.1007/978-3-319-94950-5_7

to improve on humankind, regardless of whether or not any pathology needing correction is present.

Similarly, The Center for Cognitive Liberty and Ethics, which maintains an online presence at http://www.cognitiveliberty.org, is a network of scholars debating the law, policy and ethics of freedom of thought with a mission "to develop social policies that will preserve and enhance freedom of thought into the 21st century." They argue that the "definitions of 'medicine' and 'mental health' are expanding from treatment and prevention, to improvement and enhancement" and that the capabilities of the human mind should not be limited: "So long as people do not directly harm others, governments should not criminally prohibit cognitive enhancement or the experience of any mental state" (http://www.cognitiveliberty.org/mission.html).

7.2 Transhumanism as Fringe Science

One concern that has arisen among commentators discussing transhumanism is that many aspects of transhumanism constitute "fringe science", defined here as scientific investigation in a traditional field of inquiry (e.g., biology) that departs significantly from accepted theories, and consequently lies in the "fringes" of a mainstream academic discipline. One example of a transhumanist domain that is sometimes accused of existing in the fringes is Aubrey de Grey's anti-senescence program, which is based on the notion that a scientific program aimed at mitigating seven identified types of biological aging processes could extend the human life-span by a great many decades (De Grey 1997). However, as science progresses over time from relative ignorance to a more comprehensive understanding it necessarily passes through a transition phase where confusion and uncertainty predominate. This would be the case, for instance, for the beginnings of evolutionary theory in biology and for the early years of quantum mechanics.

7.3 Ethical Concerns

While many individuals dismiss transhumanist ideas as mere impractical fringe science, a number of opponents to transhumanism offer ethical concerns as the basis for their opposition, such as a fear that human enhancement technologies would debase us and undermine our human dignity, or concerns that such technologies would be deployed unfairly, benefiting only the wealthy. With respect to the latter issue, Francis Fukuyama has expressed a fear that deploying transhumanist approaches to improving the human condition necessarily redirects needed resources from implementing (arguably more practical) social solutions (Fukuyama 2004).

Other critics advance the notion that it is fundamentally wrong to alter humanity and human nature in the first place. For instance, Leon Kass famously argues that we should avoid these technologies because they are repugnant (Kass 1997), while other "bioconservatives" such as Michael Sandel, George Annas, Jeremy Rifkin and Wesley Smith all offer related reasons why human enhancement technologies are, in their opinion, ethically suspect.[1]

Certainly, some of the scenarios described in earlier sections of this book raise troubling (or at least interesting) ethical questions. If the aim of the medical profession is to enhance the health and the quality of life of individuals who are sick, is it ethical that medical knowledge be used to improve the lives of folks who are already quite healthy? Will university faculties of education eventually have departments of educational pharmacology to help advance learning enhancement by chemical means? Will medical students be allowed to use pharmacologic assistance in order to prepare for their pharmacology finals, or will they perhaps instead be required to provide a post-test urine sample to establish the opposite?[2]

There is also the practical question about the potential for long-term harm for some proposed human enhancement methods that rely on pharmacologic methods. While many individuals would like to think faster and more clearly as well as hold more things in memory, and perhaps even be able to wonder about things that cannot ordinarily be conceived, they may also ask "At what cost to brain function and to life in general does this enhancement come?"

As developments in neuropsychology and neuropharmacology provide new treatments for neurological conditions like Alzheimer's disease it appears that many of these treatments may be of interest to "normal" individuals as a means of achieving "quality of life" improvements, such as becoming better at mental arithmetic or developing a better memory. But such enhancements are now associated with a new ethical debate, with proponents arguing for their use on the basis of the principle of "cognitive liberty" and with opponents arguing against their use on the basis of concerns such as safety and social fairness.

7.4 Leon Kass and the Wisdom of Repugnance

For instance, the conservative University of Chicago philosopher, Leon Kass, mentioned earlier, is an opponent of research into life extension, stem-cell techniques and other enhancement technologies (Kass 1997). One basis for his opposition is a practical concern for safety. Another concern cited above is based on the simple notion of repugnance, as exemplified in the following comments he has made on cloning: "We are repelled by the prospect of cloning human beings not because of the strangeness or novelty of the undertaking, but because we intuit and feel, immediately and without argument, the violation of things that we rightfully hold dear" (Kass 1997). That

[1] The term "bioconservative" is used here and elsewhere (Sect. 7.7) without any intended pejorative sense, reflecting a well-developed philosophical position with which, as discussed in further sections, I am in general disagreement. The term "bioluddite" is sometimes used when contempt is intended.

[2] It has not escaped my attention that university students already often use pharmacologic assistance in an attempt to earn higher marks.

is, Kass argues that notions that are repugnant to our sensibilities—for instance, the idea of making human-animal hybrids—should be viewed as being unethical on the basis of the revulsion they generate.

Not surprisingly, many people in the transhumanist community find such an argument to be weak and even preposterous, being based, they argue, on an appeal to emotion and a rejection of rationalism (Roache and Clarke 2009; Turner 2004). As a simple example, much of what is done every day in hospital operating rooms around the world is repugnant to many nonclinicians, but this does not stop surgery from being conducted. Or, for that matter, it does not stop a good number of people from enjoying steak tartar and sushi.

7.5 Francis Fukuyama

Another noted opponent of human enhancement technologies is the political economist and philosopher mentioned earlier, Francis Fukuyama. Perhaps best known for his book *The End of History and the Last Man*, Fukuyama has conservative views rather similar to those of Kass. In Fukuyama's book *Our Posthuman Future* (Fukuyama 2002) he argues for banning many human enhancement technologies. His concern focuses on the potential for social conflict, with the rich having access to technologies to make them, and their offspring, "smarter, stronger, and longer-living". This is a rather different issue than the questions raised by Kass in that Fukuyama argues that transhumanism, by fundamentally altering the character of "human nature", undermines the egalitarian ideals of liberal democracy. As I discuss later (see Sect. 7.8 in particular), I do not see Fukuyama's concern to be a sufficiently strong argument to hold back advances in transhumanism. In particular, I do not see it as a given that transhumanist technologies that alter human nature must necessarily do this at the expense of egalitarian ideals of liberal democracy, since (for example) it is logical to expect that a number of future cognitively enhanced individuals will be able to offer to all of humankind a substantial number of scientific, medical and cultural treasures.

7.6 The President's Council on Bioethics

Yet another perspective critical of human enhancement technologies is the President's Council on Bioethics, which in 2003 published its lengthy critique of human enhancement, entitled *Beyond Therapy: Biotechnology and the Pursuit of Human Improvement* (The President's Council on Bioethics 2003). The report is unusually comprehensive in its scope, variously considering the use of drugs and gene transfers to enhance augmenting muscle strength and athletic performance, the use of mind-altering psychoactive drugs, techniques for controlling the sex of fetuses, techniques intended to lengthen the human lifespan, and related issues.

Most of the discussions in the report, however, take on a clear bioconservative, cautionary tone. This is hardly a surprise, given that the President's Council on Bioethics, the overseer of the report, at the time was a politically appointed group composed almost entirely of individuals with conservative views that were closely aligned to those of the American (second) Bush administration. As a result, some critics argue, the report is lacking in balance and perhaps even in scientific accuracy. For instance, Blackburn and Rowley (2004) write: "In the discussions of preimplantation genetic diagnosis, the specter of designer babies is raised by implying that selecting embryos for intelligence and other traits, such as temperament is a possibility. Scientifically, this simply is highly unlikely and indeed may not even be feasible. While such scientific unlikelihood is mentioned in passing in the report, it is easy to take away from the report the feeling that such genetic manipulation will happen and is even imminent." Similar concerns have been raised by other individuals (Sinsheimer 2004).

7.7 Exploring Bioconservative Arguments

I would like now to add a contemplative component to the commentary by including discussion of some of my own views. In a nutshell, I have found that the more that I explore these issues, the more I ultimately find the arguments offered by the bioconservatives to be unsatisfactory and unconvincing.

In general, bioconservatives offer six broad categories of reasons why human enhancement by artificial means should not be undertaken. First, they may cite reasons of social fairness. Second, they may cite reasons related to the endangerment of our dignity or our humanity. Third, they may cite "slippery slope" issues. Fourth, they may raise a concern about repugnance for transhumanist creations. Fifth, they may argue against "Playing God." And finally, they may cite reasons related to safety.

As I discuss in later sections, I regard the first fives lines of argument to be fragile and weak, while I believe that the last argument (concerning safety) has merit.

7.8 The Fairness Issue

Consider first the fairness issue, exemplified by the arguments offered by Fukuyama (2002, 2004) who presents the case that human enhancement technologies present too much potential for social conflict. He is concerned that only the rich and the socially connected would have access to these technologies. However, this concern, in my view, is excessively speculative and unsupported by strong evidence.

First, many technologies become vastly cheaper over time, and there is no reason to believe that this would not also be the case for at least some future human enhancement technologies. Using the automobile as an historical example, would an earlier Fukuyama have banned automobiles in the era before Ford introduced mass production for the same reasons?

Secondly, it appears to me that the fact that the rich get access to better housing, better vacations, better automobiles and better schools is not in itself a good reason to require that all citizens should live at the same standard of living. Would Fukuyama also ban yachts, truffles, aesthetic surgical procedures like facelifts, and all other items that the poor generally cannot afford?

Now, an argument can be made that these examples are not things that exacerbate social and economic inequality but things that follow from it. Accepting this line of reasoning for the purposes of discussion, how might one ensure that human enhancement does not necessarily contribute to social inequality? One approach is to recognize expected "trickle down" effects, such as the cognitively enhanced medical scientist who develops a novel cure for (say) ovarian cancer, with benefits that extend to all womankind, or the cognitively enhanced writer who constructs a timeless literary masterpiece valued by countless future generations.

Another approach would be to require more substantive contributions to society from enhanced humans, reminiscent of the Guardian Class in Plato's *Republic*.[3]

Yet another line of argument recognizes the reality that human nature is already unequal (by virtue of the genetic and social lottery that takes place at conception) and argues that attempts at the equalization of human nature are simply unrealistic in view of this reality (and, arguably, in view of failed national experiments in Marxism.) Instead, it could be argued, differences in human traits, talents and capabilities should result in differences in expected contributions to society.

In summary, when Fukuyama and others of his ilk argue that transhumanism fundamentally alters the character of human nature, and thereby undermines the egalitarian ideals of liberal democracy (thus generating potential for social conflict), I find these arguments to be weak, unengaging, and ultimately inconsistent with the principle that individuals using legal and moral means should be free to accumulate as much wealth as they desire.[4]

---

[3]Plato envisioned his ideal society as consisting of three classes: the producers, the auxiliaries, and the guardians. The producers included farmers, craftsmen, merchants, artists, and so forth, and they did not participate in governing. The auxiliaries were the warrior class, responsible for defense and policing; they too did not govern. Only the guardians, picked from the ranks of the auxiliaries, were responsible for governing while living in austere conditions, as they would not own private property, would live frugal, communal lives, and would have minimal privacy. Plato envisioned that the guardians would even possess wives and children in common!

[4]I say this provided that proportionate contributions to society follow, as with the philanthropy example provided by Bill Gates. While the principle of liberal freedoms would allow individuals to accumulate wealth by legal means through hard work, enterprise, innovation and imagination, this does not suggest that these same individuals do not have a moral responsibility to shift some of their wealth to less fortunate individuals (notwithstanding the followers of Ayn Rand and her philosophy of ethical egoism). Such a redistribution of wealth is typically achieved via taxation, although some thinkers argue that wealth redistribution via philanthropy is an even more acceptable method. Of course, in this setting the issue is not so much whether such redistribution should take place, but rather the extent to which wealth should be distributed from the rich to the less fortunate.

7.9 The Dignity Issue

A second category of arguments against transhumanism that bioconservatives sometimes offer is a concern that transhumanism endangers human dignity or denatures our humanity. As an example in this category, consider the work of Michael Sandel, a Harvard University professor of government. In his recent book *The Case Against Perfection* (Sandel 2007), he presents the argument that pursuits toward attaining human perfection by artificial means should not be attempted because such activities are said to endanger our humanity, imperiling the notion of human life as a gift and eroding our appreciation for the natural talents and endowments that human life provides.

Thus, for example, according to Sandel, athletes should pursue excellence through the development of their natural gifts, rather than employ technical mastery over nature as a means to athletic prowess. Sandel writes: "The problem with eugenics and genetic engineering is that they represent the one-sided triumph of willfulness over giftedness, of domination over reverence, of molding over beholding."

However, as with Fukuyama, I find Sandel's arguments to be unpersuasive, in large part because he confuses what is natural with what is moral. What is natural, however, is not automatically what is moral, and vice versa. Nature has no moral compass; she follows no moral principles. Nature is driven by chance and the laws of physics in the case of inanimate natural phenomena (e.g., earth quakes and hurricanes), and, in the case of biological organisms, is driven by evolutionary forces as well as basic biological drives such as hunger.

Sandel offers us convictions, not arguments, and, in the end, I see Sandel's position to be mere religious teachings about the sacredness of divine gifts skillfully disguised along secular lines. In essence, Sandel seems to object to efforts in attaining human perfection by artificial means primarily because he believes that it represents a misunderstanding of our place in the Cosmos, of our relation to God. This is certainly not a secular line of reasoning. Nor is it a line of reasoning that I find to be persuasive. Consider instead the more convincing line of thinking offered by Bostrom.

In his paper, *In Defense of Posthuman Dignity* (Bostrom 2005), Bostrom argues that while bioconservatives are opposed to the use of technology to modify human nature for reasons such as the fear of producing a debased 'posthuman' product or that they might pose a threat to ordinary folk, their fears are largely unfounded as long as a concept of dignity that includes posthumans is recognized. Bostrom acknowledges that some ways of modifying human nature could be debasing, but rightly points out that obtaining technical mastery over our own nature does not inevitably lead to dehumanization. He emphasizes that when one examines what one means by the term "dignity" (the quality of being worthy or honorable; the right to be treated with respect), dignity is indeed something that a posthuman could possess. With respect to the second fear (of threats to ordinary humans) Bostrom writes:

> Human society is always at risk of some group deciding to view another group of humans as fit for slavery or slaughter. To counteract such tendencies, modern societies have created laws and institutions, and endowed them with powers of enforcement, that act to prevent groups

> of citizens from enslaving or slaughtering one another. The efficacy of these institutions does not depend on all citizens having equal capacities. Modern, peaceful societies can have large numbers of people with diminished physical or mental capacities along with many other people who may be exceptionally physically strong or healthy or intellectually talented in various ways. Adding people with technologically enhanced capacities to this already broad distribution of ability would not need to rip society apart or trigger genocide or enslavement. (Bostrom 2005:206)

7.10 The "Slippery Slope" Issue

A third concern sometimes raised by bioconservatives is the "slippery slope" issue. For instance, many critics accuse Leon Kass of falling into this trap when he presents the following concerns about offering complete reproductive freedom in the context of cloning (Kass 1997):

> The principle of reproductive freedom as currently enunciated by the proponents of cloning logically embraces the ethical acceptability of sliding down the entire rest of the slope – to producing children ectogenetically from sperm to term (should it become feasible) and to producing children whose entire genetic makeup will be the product of parental eugenic planning and choice. If reproductive freedom means the right to have a child of one's own choosing, by whatever means, it knows and accepts no limits.[5]

The slippery slope argument in philosophy is an argument in which a person asserts that some disastrous event will follow inevitably from a more limited or more benign prior event, usually without offering any substantive argument for of how one event will inescapably lead to another. Examples of the slippery slope line of argument abound in everyday life. An example:

> There should be no limits on freedom of speech. Once the government starts banning one form of speech, they will never stop. Next thing you know, they will be banning free assembly and free elections!

One of the best known and emotionally compelling arguments of this type has been put into the form of a poem by Martin Niemöller (1892–1984) about the failure of the German intellectual community to fight against the Nazi purging of their chosen targets:

> First they came for the communists, and I did not speak out–
>
> because I was not a communist;
>
> Then they came for the socialists, and I did not speak out–
>
> because I was not a socialist;
>
> Then they came for the trade unionists, and I did not speak out–
>
> because I was not a trade unionist;

[5]If asked, Kass might argue that he is not using the "slippery slope" argument in the usual way, but rather is invoking a sequence of logical steps constituting a logical inference. Kass might say that if X logically entails Y and if Y is morally unacceptable, then X should also be morally unacceptable.

Then they came for the Jews, and I did not speak out–
because I was not a Jew;
Then they came for me–
and there was no one left to speak out for me.

(Parenthetically, I have always been captivated by rather similar comments from Albert Einstein, given after World War II[6]: "When the Nazis came to power, I looked to the universities that prided themselves upon their intellectual freedom, and they failed me. I looked to the German press, which prided itself on the freedom of the press, and it failed me. Until at last the churches stood alone, and that for which I once had little regard earned my respect.").

In more formal terms the slippery slope argument goes like this.[7] If event **A** happens, then via a gradual series of small but inevitable steps events **B**, **C**, and **D** will take place, and so on, in a while leading to events **X** and **Y**, and eventually to event **Z**. But event **Z** is a terrible thing to contemplate, and should not be allowed to happen, since it is truly horrible. Therefore, event **A** should not be allowed to take place, either.

It should be emphasized that the slippery slope argument is taken to be a philosophical fallacy in those cases where no substantive argument is offered for how one event will inescapably lead to another. In the case of the Martin Niemöller poem provided above it is arguable that the poem does not constitute such a fallacy because the notion of how one event (purging the nation of Communists) would inescapably lead to another (purging the nation of Jews) was apparent to anyone familiar with Nazi ideology.

Slippery slope arguments invoke an image of an inevitable, limitless series of event relationships. Experience shows, however, that scientists and doctors are generally capable of making the distinctions needed to establish limits. For instance, morphine is a useful drug that can be deadly in large doses. But no reasonable person uses the slippery slope argument to suggest that the clinical use of morphine should be outlawed because if doctors are allowed to give 5 mg of morphine to eliminate pain, they will eventually be giving 500 mg doses to eliminate life. One obvious hard limit, for instance, would be to forbid carrying out procedures (of any kind) on competent adults against their will. Similar limit rules can also be envisioned in other settings.

It is of interest to policy analysts and lawmakers to see whether slippery slope arguments have ever been validated empirically following a change in law or policy that was predicted to have morally disastrous effects. For instance, Ryan (1998), using data from Australia and the Netherlands, examined the evidence "for the empirical argument that there is a slippery slope between the legalization of voluntary and non-voluntary euthanasia." He concluded that "contrary to the warnings of the slippery slope, the available evidence suggests that the legalization of physician-assisted suicide might actually decrease the prevalence of non-voluntary and involuntary euthanasia."

---

[6]http://storiesfortrainers.com/valuesquotes.aspx.

[7]http://www.iep.utm.edu/fallacy/.

Ryan (1998) also offers the following enlightening commentary on slippery slope arguments:

> The slippery slope is what Daniel Dennett has termed an intuition pump. Intuition pumps bypass the uncertain and exhausting path of convincing readers with careful and detailed argument in favour of a more easily travelled byway that speeds readers to a conclusion based upon their gut feeling. Intuition pumps replace argument with slogans and telling images. They are designed to convince readers of the truth of what they already suspected.

7.11 The "Repugnance" Issue

A fourth category of anti-transhumanist arguments that some bioconservatives put forward is the "repugnance" argument. This argument has been popularized by Leon Kass in his previously mentioned article "The Wisdom of Repugnance" (Kass 1997), which is principally concerned with the ethics of human cloning. Distilled to its essence, Kass argues that an intuitive response of repugnance or disgust to an entity or concept should be taken as evidence that the entity or concept is intrinsically evil or harmful. He uses this as an ethical argument against human cloning by trying to make the case that human cloning should evoke disgust:

> People are repelled by many aspects of human cloning. They recoil from the prospect of mass production of human beings, with large clones of look-alikes, compromised in their individuality, the idea of father-son or mother-daughter twins; the bizarre prospects of a woman giving birth to and rearing a genetic copy of herself, her spouse or even her deceased father or mother; the grotesqueness of conceiving a child as an exact replacement for another who has died; the utilitarian creation of embryonic genetic duplicates of oneself, to be frozen away or created when necessary in case of need for homologous tissues or organs for transplantation; the narcissism of those who would clone themselves and the arrogance of others who think they know who deserves to be cloned or which genotype any child-to-be should be thrilled to receive; the Frankensteinian hubris to create human life and increasingly to control its destiny; man playing God.

Speaking candidly, I am surprised that this argument has achieved any acceptance in intellectual circles, being based on a fallacious appeal to emotion: "It's disgusting, so it has to be wrong!" The ultimate underlying premise in Kass's position seems to be that emotional responses should trump the rule of reason.

For the sake of completeness, I should point out that there is an interesting body of scientific literature that is just starting to develop that is concerned with studying the links between disgust and moral judgment. For instance Schnall et al. (2008) performed laboratory work to show that "the role of disgust in severity of moral judgments depends on participants' sensitivity to their own bodily sensations", suggesting that gut feelings are sometimes important in moral judgment. Other studies have used advanced brain imaging methods ("functional MRI") to determine which centers in the human brain are involved in the mental processes central to moral judgment (Moll et al. 2005).

While these studies and those of others (Borg et al. 2008; Gutierrez and Giner-Sorolla 2007; Wheatley and Haidt 2005) are helpful in understanding the neurophysiology of intuitive moral judgment, they in no way support Kass's concept that repugnance is a reliable guide to determining which actions are immoral.

7.12 Playing God

The expression "Playing God" is usually taken to refer to an individual who is taking on the role of God, such as in deciding who is to live or die in dire clinical situations or in the circumstances of "lifeboat ethics." Alternately, it might be used to characterize a situation (for example) where a scientist creates a new form of life (van den Belt 2009). The phrase is often used to invoke a precautionary principle or mount a charge of human hubris, humans presumably not being allowed to claim for themselves that which is said to be the sole providence of the Divine (or the sole providence of nature, if you prefer). Those individuals who hold that only God (or nature) and not scientists and technicians should create new life forms or enhance existing life forms are explicitly stating that humans should be morally forbidden to carry out certain kinds of biological research. As an example, responding to the news that Craig Venter and his team had invented the synthetic organism *Mycoplasma laboratorium*, Pat Mooney wrote[8]: "For the first time, God has competition. Venter and his colleagues have breached a societal boundary, and the public hasn't even had a chance to debate the far-reaching social, ethical and environmental implications of synthetic life."[9]

Given that similar arguments about "Playing God" might be made when clinicians save lives that would otherwise be snuffed out by natural processes like premature birth, pneumonia or cancer, one wonders what biological interventions opponents of "Playing God" would forbid. Even something as simple as starting fires and cooking food could arguably be forbidden on the basis of somehow being "unnatural". Of interest, in the case of Craig Venter, he was wise enough to first solicit ethical advice from a panel of bioethicists and theologians (the so-called *Ethics of Genomics Group*) which, upon careful deliberation, identified no serious objections against Venter's plans and eventually provided their imprimatur (Cho et al. 1999). The basis for their position was that while the technology has some abuse potential (e.g. biological weapons), and cause us to rethink our conception of life, the committee did not see that construction of new genomes of the type in question was a violation of any fundamental moral precepts or boundaries.

[8]ETC Group (2007) Goodbye, Dolly … Hello, Synthia! J. Craig Venter Institute Seeks Monopoly Patents on the World's First-Ever Human-Made Life Form. Retrieved January 7, 2013, from http://www.etcgroup.org/content/patenting-pandora%E2%80%99s-bug-goodbye-dollyhello-synthia.

[9]Arguably, however, this would not be the first time that God has had "competition", since the same issue arose decades earlier in the 1970s with the debate on recombinant-DNA methods.

7.13 The Safety Issue

A final concern sometimes raised relates to safety, where fears are raised that new human enhancement technologies may end up causing unanticipated harm. Of the concerns raised by bioconservatives, this one resonates with me the most. The following cautionary tale will serve to illustrate my unease.

On September 17, 1999 Jesse Gelsinger, an 18-year-old boy, died when an experimental gene therapy treatment at the University of Pennsylvania went tragically wrong. Jesse suffered from ornithine transcarbamylase deficiency, the result of an unfortunate sporadic genetic mutation that made him unable to breakdown much of the ammonia produced as a natural consequence of his body's breakdown (catabolism) of dietary protein.

Jesse's doctors were hoping to treat his condition by implanting the missing healthy gene into his liver cells using a gene carrier, or vector, in this case a common virus (adenovirus) directly administered into his hepatic artery. Since some seventeen patients had been successfully treated for other genetic disorders using similar techniques, they were confident of success, or at least, of not harming Jesse. They were wrong. Jesse died unexpectedly of multi-system organ failure four days after starting the treatment, likely from a severe immunological response to the adenovirus. An inquiry followed that identified a number of problems with the clinical trial process. Jesse's father sued and subsequently settled out of court for an undisclosed amount (Stolberg 1999).

While the above tragic example is thought-provoking, it should also be emphasized that such safety concerns are hardly unique to transhumanist endeavours—they also apply to virtually all fields of science and medicine, at least to some degree. Similar issues arose in the early years of aviation, where many pilots lost their lives or, more recently, with risky medical advances such as open-heart surgery or transplantation medicine. For instance, the famous transplant surgeon Thomas E. Starzl attempted 5 human liver transplants between March 1 and October 4, 1963. The first patient bled to death during the operation while the others died after 6.5–23 days.[10] Such dismal outcomes led many to criticize his efforts, but in each case a careful review of the cause of death led to new insights that eventually made the procedure the highly successful operation it is today, with a five-year survival rate about 72%.[11]

Sometimes the scientific community makes an unusually special effort to address potential safety concerns in emerging technologies. In the early days of genetic engineering, for example, biologists were faced with a series of ethical and safety questions regarding recombinant DNA technologies not unlike some of the contemporary questions that arise in the context of human genomic enhancement. To address the potential biohazards and a possible need to regulate this biotechnology, a group of biologists, physicians and lawyers gathered together in February 1975 to draw up voluntary guidelines to ensure the safety of recombinant DNA technology. Known as the Asilomar conference, the participants called for, and observed, a series of safety

---

[10]http://www.starzl.pitt.edu/transplantation/organs/liver.html.

[11]http://www.mayoclinic.org/tests-procedures/liver-transplant/basics/results/prc-20014076.

controls on certain experiments. In addition to regulating certain experiments, the guidelines[12] also forbade the performance of some experiments perceived to be especially risky, such as the cloning of recombinant DNAs derived from highly pathogenic organisms (Berg 2008). Conferences modelled after the 1975 Asilomar event are one obvious approach to dealing with some of the safety issues that arise with human genomic enhancement and other transhumanist technologies. In fact, in 2015 scientists called for a partial ban on genome editing in human germline cells with a view to issues not unlike those identified in the original Asilomar conference (Guttinger 2017; Lanphier et al. 2015).

It should be emphasized that transhumanists themselves are generally in favor of reasonable and thoughtful restrictions on emerging technologies with a view to mitigating possible risks. Instead, their concern is with unrealistically pessimistic risk assessments often advanced by ignorant "bioluddites" holding uninformed anti-scientific views. The contemporary anti-vaccination movement in the USA and elsewhere serves as a particularly unfortunate demonstration of how well-meaning but misguided individuals may hijack the national discourse on a biotechnical issue to the detriment of society (Leask 2015).

The above comments notwithstanding, however, I still maintain that the safety issue in transhumanist endeavors remains the most pressing one remaining to be settled. I am not alone in this matter. For example, the prominent transhumanist James Hughes has argued that transhumanist technologies necessitate not only that we take a reasoned approach to their development and deployment, but also have government bodies regulate such technologies. Hughes argues that this is necessary not only for safety reasons but also to ensure that individuals are protected from any dangers such advances may present.[13] In a similar vein, the next section presents additional cautionary commentary from two prominent scientists.

7.14 Existential Risks

In recent years a number of thinkers have become concerned that scientific and technological innovations may ultimately turn out to be profoundly harmful to humanity, perhaps even leading to the demise of civilization as we know it. Of course, since the detonation of the first atomic bomb in 1945 and the subsequent cold war, the concern that humankind could annihilate itself has figured prominently in the world's psyche. Since that time, however, a number of new existential threats have been identified and debated, many of these being based, arguably, on technologies favored by enthusiasts of transhumanism. Examples include (1) the accelerated depletion of the earth's resources, with implications for global warming and environmental contamination (Keller 2003, 2007; Fiore et al. 2012); (2) the deliberate or accidental release of new viral or bacterial microorganisms with especially virulent properties, such as might be

[12]http://profiles.nlm.nih.gov/QQ/B/C/G/D/_/qqbcgd.pdf.

[13]http://hplusmagazine.com/2014/07/29/ethics-and-policy-concerns-in-the-transhuman-transition/.

constructed in bioweapons laboratories (Nuzzo 2006; Friedman et al. 2010); and (3) the so-called "grey goo scenario" where out-of-control self-replicating nanorobots consume all the earth's resources in an orgy of self-proliferation, resulting in global catastrophe (Giles 2004).

Faced with such concerns, a number of thinkers have argued that the current accelerated rate of scientific and technological discoveries poses a clear and present danger to humanity. For example, in his famous essay, *Why the Future Doesn't Need Us*[14] renowned computer scientist Bill Joy argues that human beings are likely to destroy themselves as a consequence of technological innovation. From his paper:

> Thoreau also said that we will be "rich in proportion to the number of things which we can afford to let alone." We each seek to be happy, but it would seem worthwhile to question whether we need to take such a high risk of total destruction to gain yet more knowledge and yet more things; common sense says that there is a limit to our material needs - and that certain knowledge is too dangerous and is best forgone.
>
> Neither should we pursue near immortality without considering the costs, without considering the commensurate increase in the risk of extinction. Immortality, while perhaps the original, is certainly not the only possible utopian dream.

British Astronomer Royal Martin Rees offers a similar narrative. In his book *Our Final Hour* (2003), he argues that innovations in science and technology bring as much risk of calamity as opportunities for human progress, and while not calling for an end to scientific activity, he does argue for tighter security in the dissemination of potentially dangerous scientific information, as well as for better educating the scientific community as to the potentially hazardous consequences of some forms of scientific knowledge.[15]

Such issues are the subject of continuing lively debate within the transhumanist community. As noted earlier, as a rule, transhumanists do not argue that there should be no restrictions whatever on emerging technologies but view some of the proposals circulated by bioconservatives and their fellow travelers as unrealistic and unduly restrictive, potentially depriving humanity of timely access to badly needed safe, clean technologies.

Governments have two principal ways to limit scientific initiatives. First, they can limit the government grant funding needed to support academic research, as the US government under President George W Bush did to discourage embryonic stem cell research (Spiegel 2013). Secondly, they can introduce legislation and regulations to explicitly forbid certain lines of research, regardless of where the needed funding is obtained. An approach proposed by Bostrom (2002) is to differentially influence

---

[14]http://www.wired.com/wired/archive/8.04/joy_pr.html.

[15]A possible example is the publication in the journal, *Science*, of information concerning the viral genome of the 1918 Spanish influenza pandemic virus, which killed as many as 50 million people worldwide, because of fears that this information could be used for evil purposes (Tumpey et al. 2005).The rationale for this work was altruistic: by reconstructing the virus, researchers learned which genes were responsible for making the virus so dangerous, which in turn helps in the development of new drugs and vaccines. For an interesting discussion on this matter, see http://www.readcube.com/articles/10.1038/sj.hdy.6800911.

technological development, a process where society would retard the progress of technologies thought to be possibly harmful, while accelerating the development of technologies expected to be beneficial. This might be carried out, for example, through the thoughtful allocation of research funding as well as by regulatory approval. One problem with this approach, of course, is the difficulty of predicting the risks and benefits of nascent technologies.

In the final analysis, however, just as generals in the military must accept laws, policies and regulations from their civilian overseers that they may disagree with (perhaps because they lack a sound basis in military science) so too scientists may find themselves restricted in the kinds of experiments that they are able to carry out by virtue of laws and policies that may be based more on prejudice or ignorance then on solid scientific evidence. In such cases it is the social responsibility of the scientific community to educate the public as to why their work is beneficial and worthwhile.

7.15 Conclusion

Based on the critiques of transhumanism offered by critics such as Leon Kass, Francis Fukuyama, and Michael Sandel, this chapter critically assessed five commonly offered philosophical arguments why human enhancement by artificial means is undesirable. These include the matter of fairness, human dignity issues, "slippery slope" issues, an argument based on "repugnance", and safety issues. It is shown that these arguments are for the most part weak and unpersuasive, although I so argue that concerns about the safety of some transhumanist endeavors is appropriate.

Chapter 8
Transhumanism and Religion

Religious people generally want to overcome the limitations of the body and live forever, just like transhumanists.

John G. Messerly (https://reasonandmeaning.com/2015/01/18/transhumanism-and-religion/)

Abstract This chapter is concerned with the intersection of transhumanism, neurology and religious belief. The role of various brain structures in religious experiences (especially the frontal lobes) is discussed, including a discussion of how pathological brain conditions such as epilepsy can influence religious beliefs. The chapter begins with a discusuion of how various thinkers such as Marx and Freud imagined that the influence of religion would weaken over time and continues on to discuss the evolutionary origins of religious belief, the nature of religious experience as well as neurological and pharmacological circumstances that may influence a person's religious experiences, including hyper-religiosity syndromes and the central role of the frontal lobes in religiosity.

8.1 Introduction and Motivation

Religion has been a particularly important aspect of almost all cultures from the beginning of mankind (Atran and Ginges 2012; Peoples and Marlowe 2012; Baumard and Boyer 2013; Boyer 2003). Even today, while formal religious belief has declined in Europe in recent years, it remains an important cultural force in much of the world, and especially in the US.

D. J. Doyle, *What Does it Mean to be Human? Life, Death, Personhood and the Transhumanist Movement*, Anticipation Science 3,
https://doi.org/10.1007/978-3-319-94950-5_8

There are many reasons why understanding the psychological and neurologic underpinnings of religion is of interest. For example, individuals who study religious cults[1] may suspect that some cult members are "brainwashed" and may wish to understand the psychological processes involved, either to assist in the process of "deprogramming" rescued individuals, or, as in the case of CIA psychologists (and those of other intelligence agencies) to develop brainwashing programs of their own.

Atheists may wish to develop a mechanistic understanding of the religious experience to support arguments that religious belief is a mere artifact of brain function (Adelman 2007; Palmer 2006; Bhavsar and Bhugra 2008; Inzlicht et al. 2009; Boyer 2003). Believers, on the other hand, may wish to argue the exact opposite based on the exact same empirical findings.

Psychiatrists, psychologists and neurologists may be interested in exploring hyper-religiosity syndromes and other phenomena with a view to determining what neurotransmitters, neural pathways and brain structures, if any, are involved in religious belief or in the experience of "religious ecstasy" as described by William James in his classic 1902 book *The Varieties of Religious Experience* as well as by more recent authors (Bhavsar and Bhugra 2008; Buckley 1981; Parnas and Henriksen 2016; Ng 2007; Beauregard and Paquette 2008). This field is sometimes known as "neurotheology".[2]

Related to the above is a similar quest to understand the brain mechanisms underlying moral or ethical decision making, a field sometimes known as "neuroethics" (Levy 2011; Kolber 2011; O'Connell 2011; Whitehouse 2011; Garnett et al. 2011). For example, in this field individual being studied may be given an ethical dilemma to consider (e.g., Is it OK to kill one innocent man to save the lives of one thousand others?). In this scenario, real-time metabolic brain imaging (e.g. functional-MRI methods) might be carried out to determine which brain structures are involved in thinking through such an issue.

Finally, Charlton (2008) has argued that genetic engineering might be employed by some future clinician-scientists for "genospiritual enhancement" with a view to making life more "meaningful." Charlton calls this fascinating possibility "genospiritual engineering" and is an example of a proposed transhumanist enhancement with clear religious implications. (Genospiritual engineering is discussed in more detail in Sect. 8.16.)

These and related themes are the subject of this chapter, which is intended as an all-too-brief overview of the winding yet intersecting roads of religion, psychology and ethics in the context of neurology and transhumanism. Those readers who thirst

---

[1]Of course, the distinction between a cult and a religion is arguably only a matter of time, size and perspective. On this matter, the interested reader is directed to an enlightening Wall Street Journal article entitled "When Does a Religion Become a Cult?, available online at http://online.wsj.com/article/SB10001424052748703529004576160363526909654.html.

[2]For two excellent introductory accounts of neurotheology the interested reader is directed to the following recent books: *The "God" Part of the Brain: A Scientific Interpretation of Human Spirituality and God*, by Matthew Alper (http://www.amazon.com/God-Part-Brain-Interpretation-Spirituality/dp/1402214529) and *The God Gene: How Faith Is Hardwired into Our Genes*, by Dean Hamer (http://www.amazon.com/God-Gene-Faith-Hardwired-Genes/dp/0385720319).

for more details (admittedly, the present account only lightly scratches the surface of a complex yet absorbing topic full of philosophical side roads) are directed to the references listed throughout, where many specialized treatments of the subject are identified.

8.2 Religion and Modernity: Views from Marx, Freud, Weber, Einstein, Harris and Dawkins

A number of thinkers such as Marx, Weber and Freud imagined that with the advent of modernity, the influence of religion on society would diminish.

Marx

Marx, focusing on the notion of social alienation produced by the capitalist system, argued that religion, which sought to somehow sooth this alienation with promises of a better life in the afterworld, would vanish as the capitalist system became replaced by a communist system in a process he described as "historically inevitable" (Marx and Engels 2002).

Weber

Weber offered a socioevolutionary model of religious progression, arguing that, as a general rule, societies progress successively from primitive magic through the stages of polytheism, pantheism, monotheism and finally, a system focusing on secular ethical principles (Weber 1933).

Freud

Freud saw God as an illusion based on man's infantile need for a powerful father figure. He similarly saw religion as an evolutionary adaptation that served to help individuals restrain their violent impulses. Freud argued that as civilization progresses, religious beliefs would be set aside, replaced by reason, analysis and science (Capps 2001).

In all the above cases, the predictions made did not come true in America. Certainly, classical communism has not taken over America or any significant portion of the world. Nor have religious belief systems been universally set aside in favor of a more scientific approach, as Freud expected, at least not in America, although this is largely true in Europe.

In more recent times thinkers such as Einstein, Harris and Dawkins have critically examined the neurological and psychological underpinnings of religious belief to offer additional insights.

Einstein

In a 1930 New York Times article, Albert Einstein distinguished three human characteristics which lead to religious belief: fear, a human need for social morality, and an appreciation of cosmic awe.[3] Each of these impulses leads to a different form of religious belief system.

In primitive societies an unsophisticated appreciation of the physical nature of causality leads to fear, with the invention of supernatural beings who, when appropriately paid homage, offer humans fertility, good harvests etc., while the human need for love, justice and harmony create a different path to belief in a supreme being. Both these forms of religious belief system present us with an anthropomorphic concept of God.

However, Einstein saw the third form of religious belief system, originating from a profound sense of awe and wonder, as the most mature form, noting that in this instance the individual feels "the sublimity and marvelous order which reveal themselves in nature and in the world of thought."

Einstein saw science as an antagonist of the first two styles of religious belief, but as a partner in the third. He writes:

> When one views the matter historically, one is inclined to look upon science and religion as irreconcilable antagonists, and for a very obvious reason. The man who is thoroughly convinced of the universal operation of the law of causation cannot for a moment entertain the idea of a being who interferes in the course of events—provided, of course, that he takes the hypothesis of causality really seriously. He has no use for the religion of fear and equally little for social or moral religion.

and

> I maintain that the cosmic religious feeling is the strongest and noblest motive for scientific research. Only those who realize the immense efforts and, above all, the devotion without which pioneer work in theoretical science cannot be achieved are able to grasp the strength of the emotion out of which alone such work, remote as it is from the immediate realities of life, can issue.

Harris

Samuel Benjamin "Sam" Harris is the co-founder (with his wife) and chief executive of *Project Reason*, an organization that promotes science and secularism. Trained in philosophy and neuroscience, Harris has written a number of best-selling books that are deeply critical of organized religion. These include *The End of Faith* (2004), *Letter to a Christian Nation* (2006), *The Moral Landscape: How Science Can Determine Human Values* (2010), *Free Will* (2012), *Waking Up: A Guide to Spirituality Without Religion* (2014) and *Islam and the Future of Tolerance: A Dialogue* (2015) (co-authored with Maajid Nawaz).

---

[3]Albert Einstein. Science and Religion. New York Times Magazine, November 9, 1930, pp. 1–4. Reprinted in Ideas and Opinions, Crown Publishers, Inc. 1954, pp 36–40 and in Einstein's book The World as I See It, Philosophical Library, New York, 1949, pp. 24–28. Also available online at http://www.sacred-texts.com/aor/einstein/einsci.htm.

In his books, speeches and media appearances, Harris makes the case for a non-threatening, even gentle, form of religious intolerance, where stated religious beliefs are weighed against available evidence in a conversational, non-adversarial dialogue centered on intellectual honesty.

In his 2010 book, *The Moral Landscape*, Harris makes the case that human well-being might best be improved via a scientific approach, and he rejects the notion that religion should determine what activities are good and honorable. He writes "Human well-being is not a random phenomenon. It depends on many factors—ranging from genetics and neurobiology to sociology and economics" and he expects that once scientists begin developing moral norms based on scientific evaluation that moral systems that draw on religion will join "astrology, witchcraft and Greek mythology on the scrapheap".[4]

Dawkins

Richard Dawkins, a British evolutionary biologist and emeritus fellow of New College, Oxford, is perhaps best known as a militant atheist. Although many of his books deal with evolutionary biology, his views on religion are best summarized in his 2006 book *The God Delusion*. In this book Dawkins proclaims that belief in God is not only irrational but has inflicted significant damage to society. Fundamentalist varieties of Christianity and most Islamic traditions come in for particular scorn from Dawkins. To Dawkins religion is a divisive and oppressive force without which the world would be better off.

Dawkins expresses puzzlement how belief in God still exists among individuals who are scientifically sophisticated. While Dawkins understands that some physicists may use the word 'God' as a metaphor for the awe-inspiring mysteries that physicists seek to unravel,[5] he emphasizes that this metaphor is entirely different from (for example) the Christian God who conducts miracles, smites evil people and forgives sins.

Dawkins disagrees with Stephen Jay Gould's famous principle of nonoverlapping magisteria, which argues that science and religion each represent different areas of inquiry, taking instead the position that the question of the existence of God should be treated as a scientific question like any other question about the world we live in. He writes[6]:

> It is completely unrealistic to claim, as Gould and many others do, that religion keeps itself away from science's turf, restricting itself to morals and values. A universe with a supernatural presence would be a fundamentally and qualitatively different kind of universe from one without. The difference is, inescapably, a scientific difference. Religions make existence claims, and this means scientific claims.

---

[4]https://www.newscientist.com/article/mg20827822-100-morality-we-can-send-religion-to-the-scrap-heap/.

[5]For example, Einstein's famous quip that "God does not play dice with the universe" was a metaphorical statement about quantum physics rather than a theological statement.

[6]Dawkins, Richard (1998). "When Religion Steps on Science's Turf". Free Inquiry. Volume 18, Number 2. Available online at http://pds4.egloos.com/pds/200709/04/59/Richard_Dawkins_-_When_Religion_Steps_On_Sciences_Turf.pdf.

8.3 The Evolutionary Origins of Religious Belief

Anthropological Perspectives

Religion has been an important part of human civilization from prehistoric times. The culturally ubiquitous nature of religion and religious practices has led to the academic subspecialties of religious anthropology and (more recently) the sociology of religion. For example, King (2007) points out that both elephants as well as some nonhuman primates such as chimpanzees and bonobos exhibit proto-religious traits such as carrying out rituals surrounding the death of community members.

Many evolutionary biologists and (especially) evolutionary psychologists have argued that the origins of religious thought lie in the evolution of genes that resulted in a brain of sufficient size and complexity to permit humans to contemplate religious and philosophical issues (Soeling and Voland 2002; Baschetti 2005; Doebeli and Ispolatov 2010; Bloom 2012). They also argue that that religion has been evolutionarily programmed into the human brain via natural selection (evolutionary origin of religions).[7] For instance, arguing that the emergence of religious thought requires a system of symbolic communication, Nicholas Wade in his 2009 book *The Faith Instinct: How Religion Evolved and Why It Endures* writes:

> Like most behaviors that are found in societies throughout the world, religion must have been present in the ancestral human population before the dispersal from Africa 50,000 years ago. Although religious rituals usually involve dance and music, they are also very verbal, since the sacred truths have to be stated. If so, religion, at least in its modern form, cannot pre-date the emergence of language. It has been argued earlier that language attained its modern state shortly before the exodus from Africa. If religion had to await the evolution of modern, articulate language, then it too would have emerged shortly before 50,000 years ago.

Arguably, if one were to look for the seat of religion in the brain drawing on anthropological and evolutionary perspectives, one would begin with the frontal lobes, as the frontal lobes are among the "newest" brain structures from an evolutionary viewpoint. Indeed, this possibility is supported both by anecdotal clinical evidence and well as from evidence from neuroimaging studies. These issues are discussed in more detail in later sections of this chapter.

The God Gene Hypothesis

The God gene hypothesis, suggested by geneticist Dean Hamer, offers us the notion that a particular gene (specifically, vesicular monoamine transporter 2—VMAT2), predisposes individuals towards enhanced spiritual/mystic/religious experiences by altering brain monoamine levels. In his 2005 book *The God Gene: How Faith is Hardwired into our Genes*, Hamer elaborates at length on this hypothesis, drawing on genetic, neurobiological and psychological studies.

---

[7] Mikhail (2007) has taken this matter even further, suggesting that a neurologically-based "universal moral grammar" may exist in humans, akin to Chomsky's universal grammar theory (Palmer 2006; Adelman 2007).

This intriguing hypothesis has met resistance at a number of levels. First, the VMAT2 gene merely codes for a molecular pump; to reduce religious faith down to a biological pump that transports neurotransmitters across cell membranes seems very reductionist. Second, even if the VMAT2 gene ultimately enhances spirituality, it tells us nothing about whether there is a God. To the atheist, should the God gene hypothesis be validated, it will merely serve to support his argument that religious belief can be reduced to mere biology. But to the believer, eventual proof of the God gene hypothesis would be seen as an ingenious method God chose to help humans come to embrace the realm of the divine.

In the end, perhaps all that can be said is that the VMAT2 gene, like the condition of temporal lobe epilepsy discussed later in this chapter, may be an important neurobiological pathway in the study of the neurobiology of religious belief.

God as an Attachment Figure

Attachment theory, founded by psychologists John Bowlby and Mary Ainsworth (Bretherton 1992) was initially based in the observation that human beings are born with an innate psychobiological drive to bond with "attachment figures". In a 1952 report to the World Health Organization on the importance of attachment,[8]Bowlby emphasized that "the infant and young child should experience a warm, intimate, and continuous relationship with his mother (or permanent mother substitute) in which both find satisfaction and enjoyment," and that failure to do this may have negative mental health consequences for the child. Subsequent to that report, Bowlby collaborated with Ainsworth and other individuals to more fully develop attachment theory (Bowlby 1969, 1976, 1988).

The insights provided by attachment theory have encouraged some thinkers to conceptualize religious belief in terms of an attachment relationship with a divine being that emerges as a consequence of neural mechanisms that developed under the influences of evolutionary psychology (Granqvist et al. 2010; Kirkpatrick 2004).

According to Granqvist et al. (2010) two developmental pathways to religion may be at play in the context of attachment theory:

> "The first is a "compensation" pathway involving distress regulation in the context of insecure attachment and past experiences of insensitive caregiving. Research suggests that religion as compensation might set in motion an "earned security" process for individuals who are insecure with respect to attachment. The second is a "correspondence" pathway based on secure attachment and past experiences with sensitive caregivers who were religious."

8.4 What Is a Religious Experience?

Central to any discussion concerning the psychological and neurological underpinnings of religious experience is a definition of the term "religious experience." As the following discussion indicates, this is not a straightforward task.

[8]Bolby (1952) Maternal Care and Mental Health, Geneva: World Health Organization. Available at http://apps.who.int/iris/handle/10665/40724.

The *Stanford Encyclopedia of Philosophy* offers the following commentary regarding religious experiences[9]:

> Reports of religious experiences reveal a variety of different kinds. Perhaps most are visual or auditory presentations (visions and auditions), but not through the physical eyes or ears. Subjects report "seeing" or "hearing," but quickly disavow any claim to seeing or hearing with bodily sense organs. Such experiences are easily to dismiss as hallucinations, but the subjects of the experience frequently claim that though it is entirely internal, like a hallucination or imagination, it is nevertheless a veridical experience...

The encyclopedia also notes that:

> Religious experiences can be characterized generally as experiences that seem to the person having them to be of some objective reality and to have some religious import. That reality can be an individual, a state of affairs, a fact, or even an absence, depending on the religious tradition the experience is a part of. A wide variety of kinds of experience fall under the general rubric of religious experience. The concept is vague, and the multiplicity of kinds of experiences that fall under it makes it difficult to capture in any general account.

and finally:

> Religious experience is also to be distinguished from mystical experience. Although there is obviously a close connection between the two, and mystical experiences are religious experiences, not all religious experiences qualify as mystical.

Some authorities classify religious experiences as direct or indirect. In the case of a direct religious experience the individual experiences a direct communication from his or her deity.

Indirect religious experiences are subtler, where participants may describe "communion with God" or "oneness with the universe." This makes the experience somewhat akin to a mystical experience and also bears some similarity to the religious experiences described by individuals under the influence of some hallucinogenic drugs like LSD. In both cases, the resulting psychological state is characterized by a reduced awareness of one's external environment along with an increased awareness of spiritual feelings (Saver and Rabin 1997; Nencini and Grant 2010). Additionally, visions, euphoria, and even frank hallucinations, perceived to be direct communication from God, may occur (Beit-Hallahmi and Argyle 1977; Persinger 1983). As an example of direct communication from God, the Old Testament tells the story of how Abraham experienced a direct communication from God, telling him to kill his son Isaac[10] (Genesis 22).

---

[9]http://plato.stanford.edu/entries/religious-experience/.

[10]In this Biblical account, taking place on Mount Moriah, God asks Abraham to sacrifice his son Isaac. Although God gives Abraham no reason for the request, Abraham agrees without argument. Fortunately for Isaac an angel stops Abraham at the last minute. This particular story is discussed at length in Richard Dawkins' book "The God Delusion" as an example of the potential danger of uncritical religious obedience. Dawkins also writes: "The God of the Old Testament is arguably the most unpleasant character in all fiction: jealous and proud of it; a petty, unjust, unforgiving control-freak; a vindictive, bloodthirsty ethnic cleanser; a misogynistic, homophobic, racist, infanticidal, genocidal, filicidal, pestilential, megalomaniacal, sadomasochistic, capriciously malevolent bully." (p. 31).

Of course, a central issue in this discussion is whether or not such religious experiences are "genuine". Notwithstanding the difficulties of defining what constitutes a "genuine" religious experience, difficulties arise as to whether questions of this sort can even be addressed, much less resolved, by the scientific method.

8.5 William James' the Varieties of Religious Experience

William James' classic book *The Varieties of Religious Experience*: *A Study in Human Nature* is based on his famous Gifford Lectures on "Natural Theology" presented at the University of Edinburgh between 1901 and 1902. Published in 1902, the book was instantly recognized as a highly-readable yet seminal contribution to the academic study of religion. Its importance is exemplified by the following closing comments from the August 1902 New York Times review of the book[11]:

> Everywhere there is a frolic welcome to the eccentricities and extravagances of the religious life. Many will question whether its more sober exhibitions would not have been more fruitful of results, but the interest and fascination of the treatment are beyond dispute, and so, too, is the sympathy to which nothing human is indifferent.

Examining the topic from historical, philosophical, neurological and psychological perspectives, James skillfully discusses issues such the nature and variety of religious experiences, hopes of an eternal reward in the afterlife, shame and repentance, fears of eternal punishment, religious conversion (e.g., Saint Paul being struck down from his horse), mysticism, saintliness, religious hallucinations (e.g., Saint Joan's voices) and much more.

What makes the book particularly special, however, was the degree to which James attempted to connect religious experiences to brain phenomena, discussing, for instance, religious experiences in epileptics (vide infra), the experiences of individuals suffering from mental illness, and individuals with religious visions experienced under the influence of hallucinogens (where James make specific reference to ergot poisoning, which produces an effect similar to LSD ingestion).

Although more than a century has passed since James' book was first published, it is still commercially available[12] and remains an influential classic.

[11] A Study of Man: The Varieties of Religious Experience. *The New York Times*. August 9, 1902.

[12] For a commercial copy see http://www.amazon.com/Varieties-Religious-Experience-William-James/dp/0684842971. A free download is also available at http://www.gutenberg.org/ebooks/621.

8.6 What Can We Learn from the Experiences of Epileptics?

Epilepsy and Religious Experiences

In both literature (e.g., Fyodor Dostoevsky's *The Idiot*[13]) and in religious tradition there appears to be an association between epilepsy and mystical/spiritual/religious experiences. Clinical observations over the past century support an association between transcendental experiences of this kind in relation to seizure (ictal) episodes, particularly in temporal lobe epilepsy (TLE).

According to Devinsky and Lai (2008) "between 0.4 and 3.1% of partial epilepsy patients had ictal religious experiences", with higher frequencies "found in systematic questionnaires versus spontaneous patient reports" and with such experiences occurring "most often in TLE patients with bilateral seizure foci." They also point out that such seizures can be a source of religious conversion.[14]

Devinsky and Lai also propose a possible neurologic mechanism. They point out that the limbic system is "often suggested as the critical site of religious experience due to the association with temporal lobe epilepsy and the emotional nature of the experiences" while neocortical areas are likely also involved, as "suggested by the presence of visual and auditory hallucinations, complex ideation during many religious experiences, and the large expanse of temporal neocortex." These concepts are elaborated upon later in this chapter.

The reader will note that earlier, the frontal lobes were offered as a possible site of religious thought, while the above account suggests that the limbic system may play an important role. The likelihood is that, as with many complex brain processes, a number of brain areas are involved, possibly reflecting a variety of types of religious experiences (e.g., intellectual versus mystical experiences).

Geschwind Syndrome

Some patients with temporal lobe epilepsy eventually develop Geschwind syndrome, named after Norman Geschwind, who described a cluster of behavioral phenomena and personality changes evident in some individuals after suffering lengthy periods of undertreated or refractory temporal lobe epilepsy (Benson 1991; Geschwind 1979; Veronelli et al. 2017). The syndrome includes five primary features, although not all

---

[13]The story, published serially between 1867 and 1869, concerns Prince Lev Nikolayevich Myshkin, a sensitive young man of noble Russian lineage, who has recently arrived in St. Petersburg after spending four years in a Swiss clinic for management of his epilepsy. Myshkin, a man of noble values (altruism, kindness, compassion, ethical sensitivity) subordinates his romantic feelings to higher, saintly ideals, but in the end cannot cope in the world of decadence and moral decay that surrounds him. He ends up returning to the Swiss clinic, the real world apparently being no place for a saint. *The Idiot* is considered one of Dostoevsky's greatest works, along with *Notes from Underground* (1864), *Crime and Punishment* (1866), *The Possessed* (1872), and *The Brothers Karamazov* (1880).

[14]Was the biblical St. Paul, who underwent a sudden religious conversion on the road to Damascus, an epileptic? Some experts believe so (Landsborough 1987; Brorson and Brewer 1988).

are present in all cases: hypergraphia (a compulsion to extensive writing), hyperreligiosity (a passionate tendency towards religious and philosophical interests), atypical (typically diminished) sexuality, circumstantiality (excessive verbal output and preoccupation with details), and an intensified mental life featuring "deepened cognitive and emotional responses" (Benson 1991). Aggressiveness may also be present (Garcia-Santibanez and Sarva 2015; Benson 1991).

Garcia-Santibanez and Sarva (2015, Benson 1991) describe a case of a 40-year-old man with Geschwind syndrome where hyperreligiosity was the only finding. The hyperreligiosity appeared after the patient unilaterally discontinued his anti-epileptic medications (topiramate 100 mg twice daily and lamotrigine 200 mg twice daily). After his anti-epileptic medications were restarted the hyperreligiosity vanished. The authors offer the following clinical narrative:

> On examination, he was uncooperative. He was constantly making religious remarks, saying "God is with me and I do not need doctors or medications." He would interpret every question asked to him as questioning his faith and at times attempted to convert doctors and staff to Islam. He believed everyone around him was preventing him from obtaining salvation.

Devinsky and Lai, referenced earlier, offer us some rather detailed insight into possible neurologic mechanisms involved in cases of religious experiences associated with epilepsy:

> Although psychologic and social factors such as stigma may contribute to religious experiences with epilepsy, a neurologic mechanism most likely plays a large role. The limbic system is also often suggested as the critical site of religious experience due to the association with temporal lobe epilepsy and the emotional nature of the experiences. Neocortical areas also may be involved, suggested by the presence of visual and auditory hallucinations, complex ideation during many religious experiences, and the large expanse of temporal neocortex. In contrast to the role of the temporal lobe in evoking religious experiences, alterations in frontal functions may contribute to increased religious interests as a personality trait. The two main forms of religious experience, the ongoing belief pattern and set of convictions (the religion of the everyday man) versus the ecstatic religious experience, may be predominantly localized to the frontal and temporal regions, respectively, of the right hemisphere.

Another account, this time historical, concerns the visionary mystic Wise-Knut, born in Norway in 1792. His contemporary biographers, who narrated his major life events in detail, tell us that he heard voices, had religious delusions, and suffered from severe epilepsy (Brodtkorb and Nakken 2015). This is yet another story substantiates the notion that epilepsy may have played an important role in the history of religions.

The God Helmet

The God Helmet is an experimental apparatus developed by Stanley Koren and Michael Persinger to study the effects of magnetic stimulation of the temporal lobes (Persinger and Healey 2002). The magnetic fields generated by the apparatus are much weaker than those used on transcranial magnetic stimulation; it instead the particular *pattern* of magnetic stimulation (modelled after limbic burst firing) that is said to be special to the apparatus design (Richards et al. 1993). Reports of users sensing God's presence (Persinger uses the more secular term "sensed presence") with this device brought considerable public attention and resulted in a number of television documentaries.[15]

As an example of the sensed presence experience, here is a description from obtained a 21-year-old woman (St-Pierre and Persinger 2006):

> I felt a presence behind me and then along the left side. When I tried to focus on its position, the presence moved. Every time I tried to sense where it was, it moved around. When it moved to the right side, I experienced a deep sense of security like I had not experienced before.

Persinger and his co-workers hypothesize that the left and right temporal lobes collaborate to create a unified self and when communication between them breaks down as a result of appropriate magnetic stimulation, the 'self' in the non-dominant hemisphere 'intrudes' into consciousness, a phenomenon Persinger calls 'inter-hemispheric intrusion'. Persinger additionally hypothesizes that this phenomenon, when induced by natural means such as meteorite showers, solar flares or various geomagnetic phenomena, may help explain mystical/spiritual experiences such as those experienced by a number of religious leaders (Derr and Persinger 1989).[16]

To some critics, however, Preminger's arguments are unconvincing. Some scientists maintain that the "double blinding" done in the experiments was not up to standard and that suggestibility was at play (Granqvist et al. 2005). Other critics accept that Persinger is likely describing a genuine biological phenomenon, but argue that whether the induced "sensed presence" appears as a ghostly apparition, or as an enhanced appreciation of another individual in the room, or as God peering over your shoulder, what is at play is not necessarily a "God center" area in the human brain that is responding to the magnetic stimulus, but more likely simply a hallucination or illusion of sorts, one induced electromagnetically rather than by the use of drugs.

8.7 What Pharmacologic Interventions Tell Us

As noted earlier, in James' book *The Varieties of Religious Experience* he described religious experiences induced by ergot poisoning following the ingestion of ergot-contaminated food. In more recent times LSD has been discussed in the context of the

---

[15]For an example, visit http://www.sciencechannel.com/tv-shows/through-the-wormhole/videos/through-the-wormhole-the-god-experience/.

[16]One popular magazine tells us Joseph Smith was visited by the angel Moroni before founding Mormonism, and when Charles Taze Russell started the Jehovah's Witnesses, powerful Leonid meteor showers were occurring (http://www.wired.com/1999/11/persinger/).

religious experiences it sometimes provides.[17] Consider, for example, the following anecdote about an LSD-related religious experience provided by Unger[18]:

> Suddenly, I could see my family handing me great love. It seemed to be pouring out of their hearts. I cried, not bitter tears, but tears of beauty and joy. A beautiful organ was playing in the background, and it seemed as if angels were singing. All of a sudden I was back in eternity. There was music and beauty. Peace and happiness, tranquility—could not possibly describe my feelings. My heart was filled with joy that was overwhelming. Just a beauty and peace that I have never known. At this point, I felt that time was thousands of years ago, thousands of years from now, and now.

Of course, as noted earlier, LSD is hardly the only drug to be linked to religious experiences. Drugs such as psilocybin (Griffiths et al. 2006) and mescaline have been part of religious ceremonies in a number of aboriginal cultures for centuries. In recent years much attention has focused on dimethyltryptamine (DMT), administered either as a pure chemical or as a component of ayahuasca, a shamanistic concoction used by some Amazonian Amerindians. Part of the reason for this attention is the claim by authors such as Rick Strassman, author of *DMT: The Spirit Molecule,* that DMT is a naturally occurring "spirit molecule" responsible for near-death and mystical experiences.

Such insights naturally lead the neurologically inclined to ask about the particular brain structures involved when drugs like LSD or DMT are administered. This is, in fact, a largely unexplored field which is expected to gradually expand as functional MRI (f-MRI) techniques and other brain imaging methods become more commonly employed for such purposes.[19]

For now, what is known is that these drugs bind to various brain receptors to achieve their effect, although the specific details remain far from being fully worked out.

Psychedelics and Hallucinogens

Contemporary research has led to an understanding that a number of psychedelic/hallucinogenic agents such as LSD[20] are agonists or partial agonists acting at brain serotonin receptors,[21] especially receptors located on apical dendrites

---

[17] See for instance http://onfaith.washingtonpost.com/onfaith/undergod/2008/05/lsd_and_the_religious_experien.html and http://www.psychedelic-library.org/pahnke3.htm.

[18] Unger S. M. The current status of psychotherapeutically-oriented LSD research in the United States. Unpublished paper read to the New York State Psychological Association on 30 April 1965. (Quoted in http://www.psychedelic-library.org/pahnke3.htm).

[19] Some critics would be quick to point out some of the potential difficulties of conducting brain imaging studies of psychedelic drugs. First, f-MRI studies are very expensive and are unlikely to be funded by either conservative government agencies or by sponsoring drug companies. Second, conservative government agencies like the American DEA would be expected to disapprove most applications to study such drugs (as is frequently the case, for example, with medical marijuana studies). Third, many people disagree on principle with the idea of studying psychedelic drugs. That being said, studies outside the USA are now being undertaken to address such questions (Carhart-Harris et al. 2016).

[20] More specifically, ((5R,8R)-(+)-lysergic acid-*N*,*N*-diethylamide (LSD)-25.

[21] More specifically, 5-hydroxytryptamine 2A receptors.

of neocortical pyramidal cells in layer V[22] (Nichols 2016). (An agonist is a drug that activates matching receptors in the brain or other organs. Full agonists activate receptors in proportion to their concentration, while partial agonists offer only partial efficacy relative to a full agonist, their effect relative to concentration diminishing at high concentrations.)

Gallimore (2015) has applied the science of "integrated information theory" (IIT), a contemporary model of consciousness, to help understand the action of psychedelic drugs at another level. He writes (p. 1):

> Using the results of functional neuroimaging data on the psychedelic state, the effects of psychedelic drugs on both the level and structure of consciousness can be explained in terms of the conceptual framework of IIT. This new IIT-based model of the psychedelic state provides an explanation for many of its phenomenological features, including unconstrained cognition, alterations in the structure and meaning of concepts and a sense of expanded awareness. This model also suggests that whilst cognitive flexibility, creativity, and imagination are enhanced during the psychedelic state, this occurs at the expense of cause-effect information, as well as degrading the brain's ability to organize, categorize, and differentiate the constituents of conscious experience. Furthermore, the model generates specific predictions that can be tested using a combination of functional imaging techniques, as has been applied to the study of levels of consciousness during anesthesia and following brain injury.

Such a detailed understanding of psychedelic/hallucinogenic agents may be helpful as clinicians continue to explore the possibility that these molecules may be therapeutically valuable. For example, Das et al. (2016) write (p. 222):

> The great value of LSD lies with its 'mystical experience' and 'self realization'. In a properly structured therapeutic setting, the patient can quickly develop a level of 'self understanding' and 'self acceptance'. With the therapist's help, he or she can clearly see the inadequacies in the value system which has underlain previous behavior. With the gain in 'openness' and altered understanding, the patient tries to improve and alter the previous wrongs.

Additionally, LSD and a number of other drugs may serve as drug models of schizophrenia (Murray et al. 2013; Steeds et al. 2015). For example, Steeds et al. (2015) note that "Determining the different molecular effects of those drugs whose pharmacological effects do and do not mimic the various aspects of schizophrenia has much to teach us concerning the pathogenesis of the illness."

8.8 Insights from Functional Neuroimaging Experiments

For some time neurophilosophers and neurotheologians have been interested in determining what brain structures were responsible to religious thoughts and feelings. As an example of work in this direction focusing on meditation, Newberg et al. (2001) measured changes in regional cerebral blood flow during meditation in eight Tibetan

---

[22] An apical dendrite is a one that emerges from the apex of a pyramidal neuron. Pyramidal neurons (pyramidal cells) are a special type of neuron found throughout the brain, and especially in the cerebral cortex.

Buddhists using a radioactive tracer in conjunction with single photon emission computed tomography (SPECT). The percentage change in blood flow to various brain structures between meditation and baseline was compared. The authors found significantly increased flow in the cingulate gyrus, inferior and orbital frontal cortex, dorsolateral prefrontal cortex, and thalamus during meditation.

Mohandas (2008) has reviewed the scientific literature dealing with the neural basis of spirituality, mediation and religiosity. He summarizes the results of these studies as follows:

> Neuroimaging studies have shown that meditation results in an activation of the prefrontal cortex, activation of the thalamus and the inhibitory thalamic reticular nucleus and a resultant functional deafferentation of the parietal lobe. The neurochemical change as a result of meditative practices involves all the major neurotransmitter systems. The neurotransmitter changes contribute to the amelioration of anxiety and depressive symptomatology and in part explain the psychotogenic property of meditation.

and

> The studies point to prefrontal activation, transient hypofrontality, increased frontal lobe and decreased parietal lobe activity and also to a deafferentation of the posterior superior parietal lobule (PSPL) in spirituality.

Andrew B. Newberg

Andrew B. Newberg, mentioned earlier, is an American neuroscientist and neurotheology researcher whose research on the development of neurotransmitter tracers for the evaluation of pathological and normal brain states has received widespread attention, especially in the context of religiosity. Newberg's early work on the neurophysiological foundations of religious and spiritual experiences used specialized brain imaging methods to study Buddhist meditators and Franciscan nuns in prayer. This work was eventually published in three books, *The Mystical Mind* (D Aquili and Newberg 1999), *Why God Won't Go Away* (Newberg et al. 2002) and *Why We Believe What We Believe* (Newberg and Waldman 2006).

While critics of Newberg's work have argued that insights from neuroimaging do not necessarily extrapolate to the broader domain of religious and spiritual phenomena, Newberg himself points out that such studies are only tools to evaluate the brain during religious and spiritual experiences. Writing at his Web site (http://www.andrewnewberg.com), he notes:

> Our research indicates that our only way of comprehending God, asking questions about God, and experiencing God is through the brain. But whether or not God exists "out there" is something that neuroscience cannot answer. For example, if we take a brain image of a person when she is looking at a picture, we will see various parts of the brain being activated, such as the visual cortex. But the brain image cannot tell us whether or not there actually is a picture "out there" or whether the person is creating the picture in her own mind. To a certain degree, we all create our own sense of reality. Getting at what is really real is the tricky part.

8.9 Religion and Psychopathology. Hyper-religiosity Syndromes

In William James' book *The Varieties of Religious Experience* the reader is provided with a number of anecdotes suggesting that neurological events may be responsible for the development of a hyperreligiosity syndrome in some individuals. For example, the earlier mentioned story of the sudden religious conversion of St. Paul during his journey to Damascus, with Paul subsequently devoting his entire life to Christianity, is one anecdote of hyperreligiosity that James provides. As noted earlier, Landsborough (1987) as well as others (Brorson and Brewer 1988) speculate that the event that changed everything for Paul may have been a form of epilepsy (vide infra).

Muramoto (2004) has developed an elaborate hypothesis/explanation for hyper-religiosity syndromes focusing on frontal lobe function; this is discussed later in this chapter, after a discussion on the importance of the frontal lobes in religious and spiritual thought.

8.10 The Central Role of the Frontal Lobes in Religiosity

If there were one set of brain structures that make humans uniquely human compared to other animals, it would be the frontal lobe structures. Neuroscientists have known of the importance of the frontal lobe structures as early as the middle 1800s as a result of the amazing story of Phineas P. Gage (1823–1860), an American railroad worker who sustained severe damage to his left frontal lobe as a consequence of having a large iron rod pass completely through his head in an explosion. Although it was amazing that Mr. Gage survived such a serious injury, equally amazing were the personality changes he underwent as a consequence of losing substantial portions of his left frontal lobe. Although he underwent a long and difficult period of recuperation he was subsequently able to carry out a relatively normal life, albeit one with substantial changes in personality. A contemporaneous account of these mental changes has been provided by Harlow[23]:

> The equilibrium or balance, so to speak, between his intellectual faculties and animal propensities, seems to have been destroyed. He is fitful, irreverent, indulging at times in the grossest profanity (which was not previously his custom), manifesting but little deference for his fellows, impatient of restraint or advice when it conflicts with his desires, at times pertinaciously obstinate, yet capricious and vacillating, devising many plans of future operations, which are no sooner arranged than they are abandoned in turn for others appearing more feasible. A child in his intellectual capacity and manifestations, he has the animal passions of a strong man. Previous to his injury, although untrained in the schools, he possessed a well-balanced mind, and was looked upon by those who knew him as a shrewd, smart businessman, very energetic and persistent in executing all his plans of operation. In this regard his mind was

[23]Harlow J.M. Recovery from the Passage of an Iron Bar through the Head. Publications of the Massachusetts Medical Society 1868;2:327–347. Available from: https://en.wikisource.org/wiki/Recovery_from_the_passage_of_an_iron_bar_through_the_head.

radically changed, so decidedly that his friends and acquaintances said he was "no longer Gage".

Although we are not specifically told how the loss of frontal lobe function impacted on Gage's religiosity, we do know from Harlow that he had become "impatient of restraint or advice" and had become "irreverent, indulging at times in the grossest profanity." Indeed, these and other contemporary observations are surprisingly in line with Muramoto's characterization of a hypofunctioning medial prefrontal cortex, discussed next.

8.11 The Muramoto Hypothesis

Muramoto, a now-retired clinical neurologist with an interest in the neurological basis of religious behavior, has advanced the hypothesis that the medial prefrontal cortex is central to human religious activity (Muramoto 2004). He points out that everyday tasks carried out by the medial prefrontal cortex, "such as error detection, compliance to social norms, self-reflection, and theory of mind" are also tasks that "are a key prerequisite to the maintenance of integrated religious activity." He also hypothesizes that pathological operation of the medial prefrontal cortex may lead to certain kinds of psychopathology in relation to religion, arguing that "hyperreligiosity may result from the hyperfunction of the medial prefrontal cortex, including rigid legalism (excessive error detection), excessive concern over one's existence (excessive self-reflection), and delusional interpretation of God's mind (excess of theory of mind)."

Evidently, Muramoto came to develop this hypothesis in the course of treating a patient, a longtime pastor, "who showed dramatic changes in his religious behavior as the earliest symptom" of what is known in the neuropsychiatric community as "frontal variant frontotemporal dementia." Muramoto summarizes his experience with this patient:

> In the earliest stage of the disease, he showed no detectable cognitive impairment, yet showed significant changes in his religious commitment. Magnetic resonance imaging in this stage revealed distinct focal atrophy in the bilateral medial prefrontal cortex around the area of the orbitofrontal cortex and the anterior cingulate cortex. There were no other lesions. The fact that the change in his religious behavior preceded the onset of other symptoms by years suggests that the medial prefrontal cortex may play a specific role in the integrity of religious behavior.

With this clinical anecdote and his hypothesis, Muramoto thus provides us with further insights into the role of the frontal lobes and other brain structures in religious thought.

8.12 The Problem of "Pathological" Beliefs

Thinkers as far back as Marx and Weber have taken an interest the social consequences of various religious belief systems (Schieman 2010). As an everyday example, Roman Catholic priests are not permitted to marry; the social costs of this restriction are at times hotly debated.[24]

However, a religious issue that is likely to capture our interest even more than the priestly celibacy debate concerns the recent case of a 62-year-old man who, in desiring a means to obtain eternal salvation, underwent male genital self-mutilation after contemplating the following biblical passage from the New Testament:

> "For there are some eunuchs, which were so born from their mother's womb; and there are some eunuchs, which were made eunuchs of men; and there be eunuchs, which have made themselves eunuchs for the kingdom of heaven's sake. He that is able to receive it, let him receive it." (Matthew 19:12)

Upon being seen in hospital after his surgical (mis)adventure, it became clear that the patient was "unaccompanied by his penis", which he had flushed down the toilet after surgical excision using a pocket knife.[25]

This case, and the many similar cases that have been reported in the medical literature (Clark 1981; Schwerkoske et al. 2012) serve to remind us that there are potentially devastating consequences to uncritical religious belief systems.

8.13 What Would an Atheist Say? What Would a Believer Say?

What does this all mean? The answer depends largely on the lens employed to address the issue. Atheists would be expected to argue that the identification of any neurologic mechanism behind religious experiences merely supports the notion that religious beliefs are a mere evolutionary adaptation, one simply made more apparent by pathophysiological entities like epileptic seizures or intoxication by drugs like LSD or peyote.

On the other hand, believers might simply state that any such identified neurological pathways are part of God's design to make His presence known.

Yet others may hold that there is simply insufficient information at present to draw any conclusions from the available data, while still others may hold that the worlds

[24]For differing viewpoints on this controversial matter, see, for instance, http://www.catholiceducation.org/articles/facts/fm0014.html and http://www.helium.com/items/871255-priests-should-be-permitted-to-marry.

[25]For further details see http://www.world-science.net/exclusives/120612_gospel and http://freethinker.co.uk/2012/06/14/doctor-doctor-ive-had-an-attack-of-matthew-1912/.

of the divine and the physical/biological will forever remain separate and mysterious realms.[26]

Finally, the logical positivists[27] would be expected to argue that any questions concerning the divine are entirely "meaningless".

8.14 Ethical Perspectives

For many clinicians, religious beliefs play central role in medical ethics, especially concerning issues such as abortion, circumcision, contraception, end-of-life care, organ donation, and human sexuality (Turner 2003; Brody and Macdonald 2013). Consequently, respect for religious beliefs can have important and profound bioethical implications. As a simple example, clinical nutritionists should be aware that some religious belief systems forbid the eating of meat. A more nuanced example, discussed next, is the case of the Jehovah's Witness faith.

There are about six million Jehovah's Witnesses world-wide. Orthodox Jehovah's Witnesses will not accept homologous or autologous whole blood, packed red blood cells, plasma, platelets and white blood cells, even when "clinically necessary". This can result in a challenging dilemma for physicians because a routine, safe, and potentially life-saving medical intervention is unacceptable to the patient. Anesthesiologists are particularly affected because they are almost always responsible for intraoperative transfusion management. In the realm of cancer treatment, the availability of platelet and leukocyte concentrates has dramatically improved the patient care by permitting higher doses of chemotherapeutic agents. However, since these are products forbidden to Jehovah's Witnesses, this situation also presents special challenges in the clinical management of the practicing Jehovah's Witness.

In the case of Jehovah's Witnesses, on the surface the central ethical dilemma is a conflict between beneficence and religious autonomy. But there is another potential dilemma for those who care for these patients: because Jehovah's Witness patients may not be informed about possible inconsistencies in church doctrine regarding medical issues, should clinicians attempt an in-depth discussion to encourage a "rational and truly autonomous" decision? Or should clinicians instead remain silent regardless of the degree to which they might find the patients' views to be irrational and misguided, in order to avoid the perception of "religious interference"? Muramoto (1998a, b, 1999) and others (Savulescu 1995; Elder 2000) have taken the

[26]Stephen Jay Gould's *Nonoverlapping Magisteria* model comes to mind here. Gould holds that science and religion each have "a legitimate magisterium, or domain of teaching authority," and these two domains do not overlap (http://en.wikipedia.org/wiki/Non-overlapping_magisteria).

[27]The logical positivists hold religious language to be void of meaning. They argue that statements like "God is merciful" is an empty claim—neither true nor false—since there's no mechanism to verify the statements truth by sense experience. Specifically, logical positivists hold that a proposition makes a genuine truth claim (is true or false) if and only if it is either empirical (testable by sense experience) [example: "Red ants contain formic acid"] or analytic (true by definition) [example: "Triangles have three sides"].

position that truly ethical physicians should inform their patients as fully as possible on all matters affecting their care, seeing any discussions centering on possible religious inconsistencies as merely an attempt to ensure that their patients have had the opportunity to be fully informed on an important life-and-death issue. By contrast Shander (1999) has taken the position that one should remain silent on religious issues and focus solely on clinical concerns, citing the difficulties of what he sees as essentially being an attempt at religious conversion. In short, Shander believes that the clinician has no business influencing a person's religious views.

While there are those who would dismiss the belief systems of Jehovah's Witnesses and similar faiths as dangerous and irrational and thus not worthy of respect by the clinical community, such attitudes are neither conducive to a healthy therapeutic relationship nor respectful of one of the pillars of medical ethics—respect for patient autonomy.

8.15 Dreams and Religion

Dreams consist of a series of images, sensations and thoughts occurring during sleep, frequently entailing personal significance and carrying emotional overtones. The content, significance and purpose of dreams remain both a topic of scientific investigation and the subject of endless philosophical (and psychoanalytical) enquiry. From personal experience we know that dreams can vary greatly in nature (adventurous, magical, bizarre, sexual etc.) involving events mostly outside the control of the dreamer.

The scientific study of dreams, known as oneirology, aims to understand how dreaming pertains to the operation of the brain, often with reference to the formation of memories, understanding the effects of neuropharmaceuticals, and better appreciating the basis for mental disorders. Oneirology is to be distinguished from religious or psychoanalytical dream interpretation in that the aim of oneirology is to scientifically study the neurological basis of dreams rather than examining various possible meanings associated with particular dreams.

Many thinkers see dreams as a connection to the unconscious mind[28] while others see dreams as having special religious significance. These connections will be explored in later sections.

Three Schools of Thought on Dreams

Individuals interested in dreams tend to subscribe to one of three different dream interpretation paradigms:

[28] "Dreams are the royal road to the unconscious."—Sigmund Freud, The Interpretation of Dreams.

School 1. Dreams are a pathway to the unconscious mind that are best explored using psychoanalytical techniques such as those developed by Freud and Jung.[29,30]
School 2. Dreams are at times messages from God, and can be sacred, being the occasional product of Divine agency.[31]
School 3. Dreams are a poorly understood neurological phenomenon for which little can be said at the moment except that psychoanalytical techniques and religious interpretations are not meaningful because they make statements about the world that are not falsifiable in the Popperian sense.[32]

Religious Belief Origins Hypothesis

One hypothesis about the origins of religious belief holds that the experience of dreaming in primitive man contributed to the development of religion. The 19th century British anthropologist E. B. Tylor,[33] expressed this view in his 1871 work *Primitive Culture*[34]:

> The evidence of visions corresponds with the evidence of dreams in their bearing on primitive theories of the soul, and the two classes of phenomena substantiate and supplement one another….That this soul should be looked on as surviving beyond death is a matter scarcely needing elaborate argument. Plain experience is there to teach it to every savage; his friend

---

[29]*From Wikipedia*: Psychoanalytic dream interpretation is a subdivision of dream interpretation as well as a subdivision of psychoanalysis pioneered by Sigmund Freud in the early twentieth century. Psychoanalytic dream interpretation is the process of explaining the meaning of the way the unconscious thoughts and emotions are processed in the mind during sleep. There have been multiple methods used in psychoanalytic dream interpretation, including Freud's method of dream interpretation, the symbolic method, and the decoding method. The Freudian method is the most prominently used in psychoanalysis and has been for the last century. Psychoanalytic dream interpretation is used mainly for therapeutic purposes in a variety of settings. Although these theories are used, none have been solidly proven and much has been left open to debate among researchers. Some studies have shown that areas of dream interpretation can be invalid and therefore a decline in importance has been seen in psychoanalytic dream interpretation." (https://en.wikipedia.org/wiki/Psychoanalytic_dream_interpretation).

[30]For additional commentary see Revonsuo A. The reinterpretation of dreams: an evolutionary hypothesis of the function of dreaming. Behav Brain Sci. 2000 Dec; 23(6):877–901; discussion 904-1121 as well as Kramer M. The biology of dream formation: a review and critique. J Am Acad Psychoanal. 2002 Winter; 30(4):657–71 and Valli K, Revonsuo A. The threat simulation theory in light of recent empirical evidence: a review. Am J Psychol. 2009 Spring;122(1):17–38.

[31]*From the Catholic Encyclopedia*: "Modern theologians, whilst profiting by the progress of psychological research, continue to admit the possibility of dreams supernatural in their origin, and consequently the possibility of dream-interpretation depending on supernatural communications." *Souvay, C. (1909). Interpretation of Dreams. In The Catholic Encyclopedia. New York: Robert Appleton Company. Retrieved August 24, 2016 from New Advent:* http://www.newadvent.org/cathen/05154a.htm.

[32]*From Wikipedia:* Falsifiability or refutability of a statement, hypothesis, or theory is the inherent possibility that it can be proved false. A statement is called falsifiable if it is possible to conceive of an observation or an argument which negates the statement in question. https://en.wikipedia.org/wiki/Falsifiability.

[33]https://en.wikipedia.org/wiki/Edward_Burnett_Tylor.

[34]Digitized book available at https://books.google.ae/books?id=RUMBAAAAQAAJ.

or his enemy is dead, yet still in dream or open vision he sees the spectral form which is to his philosophy a real objective being, carrying personality as it carries likeness.[35]

Similarly, Friedrich Nietzsche in his 1880 work *Human, All-Too Human*, wrote:

Misunderstanding of the dream. The man of the ages of barbarous primordial culture believed that in the dream he was getting to know a second real world: here is the origin of all metaphysics. Without the dream one would have had no occasion to divide the world into two. The dissection into soul and body is also connected with the oldest idea of the dream, likewise the postulation of a life of the soul, thus the origin of all belief in spirits, and probably also of the belief in gods. 'The dead live on, for they appear to the living in dreams': that was the conclusion one formerly drew, throughout many millennia.[36]

In Judeo-Christian theology dreams figure prominently in the Old Testament[37] as well as in the New Testament.[38] However, sometimes the reader is also explicitly warned of false or misleading dreams.[39]

Kelly Bulkeley's book *Big Dreams: The Science of Dreaming and the Origins of Religion* (Bulkeley 2016) is concerned with a special type of dream—highly memorable dreams (Bulkeley calls them "Big Dreams") that leave a lasting impact on our psyche. He offers the reader a hypothesis about the purpose and significance of Big Dreams: Big Dreams form a primal source of religious experience and arise from hard-wired brain circuits present in all neurologically complete humans.

As noted above, whether dreams are meaningful remains a central question still being hotly debated.[40] A central issue is whether dreaming is a random by-product of REM sleep physiology that does not serve any purpose. Revonsuo[41] offers us the following discussion on this matter:

The form and content of dreams is not random but organized and selective: during dreaming, the brain constructs a complex model of the world in which certain types of elements, when compared to waking life, are underrepresented whereas others are over represented. Furthermore, dream content is consistently and powerfully modulated by certain types of waking experiences.

---

[35]https://www.psychologytoday.com/blog/dreaming-in-the-digital-age/201501/the-origins-religion-in-dreaming.

[36]https://www.psychologytoday.com/blog/dreaming-in-the-digital-age/201501/the-origins-religion-in-dreaming.

[37]Examples include [1] Numbers 12:6: "He said, 'Hear now My words: If there is a prophet among you, I, the Lord, shall make Myself known to him in a vision. I shall speak with him in a dream." and [2] Daniel 1:17: "As for these four youths, God gave them learning and skill in all literature and wisdom, and Daniel had understanding in all visions and dreams."

[38]For example, Matthew 2:12: "And having been warned in a dream not to go back to Herod, they returned to their country by another route."

[39]For example, Jeremiah 23:25: "I have heard what the prophets have said who prophesy falsely in My name, saying, 'I had a dream, I had a dream!'".

[40]In a study by Morewedge and Norton (2009) the authors investigated laypeople's interpretation of their dreams and found that participants believed that their dreams contained "hidden truths" and that dreams provided "more meaningful information about the world than similar waking thoughts."

[41]Revonsuo A. The reinterpretation of dreams: an evolutionary hypothesis of the function of dreaming. Behav Brain Sci. 2000 Dec; 23(6):877–901; discussion 904–1121. Review. PubMed PMID: 11515147.

Revonsuo further suggests that the biological purpose of dreaming is to "simulate threatening events, and to rehearse threat perception and threat avoidance." He writes:

> In the ancestral environment human life was short and full of threats. Any behavioral advantage in dealing with highly dangerous events would have increased the probability of reproductive success. A dream-production mechanism that tends to select threatening waking events and simulate them over and over again in various combinations would have been valuable for the development and maintenance of threat-avoidance skills. Empirical evidence from normative dream content, children's dreams, recurrent dreams, nightmares, post traumatic dreams, and the dreams of hunter-gatherers indicates that our dream-production mechanisms are in fact specialized in the simulation of threatening events, and thus provides support to the threat simulation hypothesis of the function of dreaming.

Another model for dreams comes from Hobson and McCarley, who proposed the "the activation-synthesis model of dream production."[42] They summarize their model as follows:

> ... dreaming sleep is physiologically determined and shaped by a brain stem neuronal mechanism that can be modeled physiologically and mathematically. Formal features of the generator processes with strong implications for dream theory include periodicity and automaticity of forebrain activation, suggesting a preprogrammed neural basis for dream mentation in sleep; intense and sporadic activation of brain stem sensorimotor circuits including reticular, oculomotor, and vestibular neurons, possibly determining spatiotemporal aspects of dream imagery; and shifts in transmitter ratios, possibly accounting for dream amnesia. ... the automatically activated forebrain synthesizes the dream by comparing information generated in specific brain stem circuits with information stored in memory.

Hobson and McCarley also note:

> The primary motivating force for dreaming is not psychological but physiological since the time of occurrence and duration of dreaming sleep are quite constant, suggesting a preprogrammed, neurally determined genesis ... [I]t casts serious doubt upon the exclusively psychological significance attached to both the occurrence and quality of dreams. (p. 954)

Freudian Dream Analysis

Freud's book *The Interpretation of Dreams* (1900) saw dreams as the "royal road to the unconscious", offering accessibility to the unconscious mind not available by ordinary means. According to Freudian psychoanalytic theory, dreams are customarily driven by a need for "wish fulfillment", representing an attempt by the unconscious mind to resolve a past or present conflict. The content in dreams, however, are often not what they appear to be, and so must be interpreted using psychoanalytic methods.

In Freudian dream analysis, two types of content are distinguished: the "manifest content" (the images, thoughts, and content remembered upon awakening) and the "latent content" (the hidden meaning of the dream that must be extracted using psychanalytic methods such as free association).

---

[42]Hobson J. A., McCarley R. W. The brain as a dream state generator: an activation-synthesis hypothesis of the dream process. Am J Psychiatry. 1977 Dec; 134(12):1335-48. PubMed PMID: 21570. See also http://citeseerx.ist.psu.edu/viewdoc/download;jsessionid=6065AD2B7A81E7B62 17E1FA7D223D0E1?doi=10.1.1.308.7522&rep=rep1&type=pdf.

An example from Freud himself may help here. On the night of July 23, 1895 he experienced a dream that he called "Irma's Injection"[43]:

> A large hall - numerous guests, whom we were receiving. - Among them was Irma. I at once took her to one side, as though to answer her letter and to reproach her for not having accepted my 'solution' yet. I said to her: 'If you still get pains, it's really only your fault.' She replies: 'If you only knew what pains I've got now in my throat and stomach and abdomen - it's choking me.' - I was alarmed and looked at her. She looked pale and puffy. I thought to myself that after all I must be missing some organic trouble. I took her to the window and looked down her throat, and she showed signs of recalcitrance, like women with artificial dentures. I thought to myself that there was really no need for her to do that. - She then opened her mouth properly and on the right I found a big white patch; at another place I saw extensive whitish grey scabs upon some remarkable curly structures which were evidently modelled on the turbinal bones of the nose. - I at once called in Dr M, and he repeated the examination and confirmed it ... Dr M looked quite different from usual; he was very pale, he walked with a limp and his chin was clean-shaven ... My friend Otto was now standing beside her as well, and my friend Leopold was percussing her through her bodice and saying: 'She has a dull area low down on the left.' He also indicated that a portion of the skin on her left shoulder was infiltrated. (I noticed this, just as he did, in spite of her dress.) ... M said: 'There's no doubt it's an infection, but no matter; dysentery will supervene and the toxin will be eliminated. ... We were directly aware, too, of the origin of the infection. Not long before, when she was feeling unwell, my friend Otto had given her an injection of a preparation of propyl, propyls ... propionic acid ... trimethylamin (and I saw before me the formula for this printed in heavy type) ... Injections of this sort ought not to be given so thoughtlessly ... And probably the syringe had not been clean.

In the lengthy discussion that follows this account, Freud provided detailed associations for each part of the dream and concluded that the dream was an unconscious wish to absolve himself of blame for the unsatisfactory outcome in his treatment of a patient with nose bleeds.[44]

Jungian Dream Analysis

Swiss psychoanalyst Carl Jung was a disciple of Freud but in time developed his own ideas about how dreams are formed and how they should be interpreted. Jung believed that Freud's idea that dreams represented unfulfilled wishes was too limited an explanation and suggested that dreams reflected the fullness and complexity of the entire unconscious mind, both personal and collective.[45] Jung saw the psyche as a

---

[43]Sigmund Freud (1900). The Interpretation of Dreams.

[44]In March 1895, Freud had treated Emma Eckstein for "hysterical" nose bleeds but without a good outcome. Freud had asked Wilhelm Fliess, an otolaryngologist and close friend, to examine her to see if there was a somatic basis to her illness. Fliess, interested in psychanalytic methods, had hypothesized that the nasal bones and the female sexual organs were somehow connected and that her symptoms, possibly arising from masturbation, could be cured through nasal surgery. Fliess removed a turbinate bone from her nose, but she later badly hemorrhaged from infection (antibiotics would only come to medicine many decades later). Freud then consulted another surgeon, who discovered and removed the surgical gauze that Fliess had accidently left behind. Eckstein was left disfigured as a result of Fliess's botched surgery.

[45]*From Wikipedia:* The collective unconscious, a term coined by Carl Jung, refers to structures of the unconscious mind which are shared among beings of the same species. According to Jung, the

self-regulating entity in which conscious attitudes and outlooks were unconsciously moderated by their opposites during the dream state.

Jung offered two approaches to interpreting dreams. In the objective approach to dream interpretation, every individual in the dream refers to the person they actually are (e.g., your sister appearing in a dream is actually your sister) while in the subjective approach, every individual in the dream instead represents an aspect of the dreamer. Thus, if the dreamer is being chased by a demented man with an axe, under the subjective approach the dreamer may eventually come to recognize his own homicidal instincts, while in the objective approach the crazed man would represent a particular real individual.

Critiques of Psychoanalytical Dream Analysis

As noted earlier, one school of thought concerning dreams holds that dreams are neurological phenomena for which little can be said at the moment except that psychoanalytical techniques and religious interpretations are not meaningful because they make statements about the world that are not falsifiable in the Popperian sense. Another offered critique is that psychoanalytic dream interpretation is derived from case studies, the results of which are difficult to generalize. Yet another criticism is that elements (images, themes, thoughts, etc.) in dreams will not mean the same thing to different people, which is another way of recognizing the generalizability problem. Other concerns are that most dreams are forgotten by the time a person gets out of bed and that dream characteristics are affected by alcohol and other drugs.

Perhaps, then, all that can be really said about dreams is that they are a neurological enigma, or as Pagel (2012) states in a more detailed manner: "Whether studied from behavioral, neuroanatomical, neurochemical, pathophysiological or electrophysiological perspectives, dreaming reveals itself to be a complex cognitive state affected by a wide variety of medical, psychological, sleep and social variables."

8.16 Genospiritual Engineering

As mentioned in Sect. 8.1, genospiritual engineering is the (proposed) use of genetic engineering methods and related technologies to enhance human spiritual and religious experiences, including the possibility of increasing ethical sensitivity in humans. Charlton (2008) offers the following explanation:

> If technologies are devised which can conveniently and safely engineer genes causal of spiritual and religious behaviours, then people may become able to choose their degree of religiosity or spiritual sensitivity. For instance, it may become possible to increase the likelihood of direct religious experience - i.e. 'revelation': the subjective experience of communication from the deity. Or, people may be able to engineer 'animistic' thinking, a mode

---

human collective unconscious is populated by instincts and by archetypes: universal symbols such as the Great Mother, the Wise Old Man, the Shadow, the Tower, Water, the Tree of Life, and many more. https://en.wikipedia.org/wiki/Collective_unconscious.

> of cognition in which the significant features of the world - such as large animals, trees, distinctive landscape features - are regarded as sentient and intentional beings; so that the individual experiences a personal relationship with the world. Another potentially popular spiritual ability would probably be shamanism; in which states of altered consciousness (e.g. trances, delirium or dreams) are induced and the shaman may undergo the experience of transformations, 'soul journeys' and contact with a spirit realm. Ideally, shamanistic consciousness could be modulated such that trances were self-induced only when wanted and when it was safe and convenient; and then switched-off again completely when full alertness and concentration are necessary.

Charlton suggests that spirituality enhancing technologies might be made available to individuals who lack "the kind of spiritual or religious experiences which they wanted (or perhaps even needed) in order to lead the kind of life to which they aspired."

One approach to this possibility (and it is only a distant possibility at this point) would be to look for "genospitituality polymorphisms" in the manner similar to contemporary investigators studying polymorphisms in adrenergic and dopaminergic genes.[46]

8.17 Transhumanism and Chritianity

Various branches of Christianity vary in their stance on Transhumanism. At one extreme are the Christian Fundamentalists, most who take a strict literal interpretation of the Holy Bible, which they hold to be historically accurate in all respects. [47] Most Evangelical Christians also take a similar stance on the Holy Bible being the central

[46]Additionally, insights into individual differences in spirituality might be obtained by other methods. For instance, in an investigation of young adults with respect to individual differences in political view points Kanai et al. (2011) "found that greater liberalism was associated with increased gray matter volume in the anterior cingulate cortex, whereas greater conservatism was associated with increased volume of the right amygdala." The authors suggest that this observation points to "a possible link between brain structure and psychological mechanisms that mediate political attitudes." Similarly, Blum et al. (2012) noted that "political discussions on TV, radio, or other media may be morphed by one's reward gene polymorphisms and as such, may explain the prevalence of generations of die-hard republicans and equally entrenched democratic legacies." Blum et al. also note that "voting, voting turnout and attachment to a particular political ideology is differentially related to various reward genes (e.g., 5HTT, MOA, DRD2 and DRD4)."

[47]Not withstanding the belief of fundamentalists that the Holy Bible cannot contain contradictions or inaccuracies (being the direct word of God), the following biblical passages may cause others to hold that many passages are not meant to be interpreted literally: [1] Thou shalt not kill—Exodus 20:13 versus Thus sayeth the Lord God of Israel, Put every man his sword by his side and slay every man his brother…companion…neighbor—Exodus 32:27 [2] And in the fifth month, on the seventh day of the month…Nebuzaradan…came…unto Jerusalem…—2 Kings 25:8 versus …in the fifth month, in the tenth day of the month…came Nebuzaradan…into Jerusalem…—Jeremiah 52:12 [3] For by grace are ye saved through faith…not of works—Ephesians 2:8,9 versus Ye see then how that by works a man is justified, and not by faith only—James 2:24.

authority on all matters, whether religious or secular.[48] Based on this belief, many fundamentalist and evangelical scholars of Christianity take what is essentially a bioconservative posture on Transhumanism, primarily based on their study of the Holy Bible, but also frequently reflecting an antiscience bias based on their steadfast belief that the Holy Bible is must be taken as literal truth and cannot contain errors or contradictions.

As an example documenting the tendency of some Christian factions to dismiss scientific accounts, consider the comments of Sarah Kroop Brown, National Association of Evangelicals Communications Director and Editor of *Evangelicals* magazine: "Evangelicals are more than twice as likely as the general public (29% vs. 14%) to say that science and religion are in conflict and that they are on the side of religion. Distrust of scientists—their motives, expertise and influence—and the role of science in society underlie much of the tension."[49]

Another example of this anti-science tendency is exemplified in variations of the fundamentalist slogan "The Holy Bible says it, I believe it, so that settles the issue."[50] For instance, the fundamentalist organization ***Answers in Genesis*** (https://answersingenesis.org) is an "apologetics ministry" that has been particularly active in "providing answers to questions about the Bible—particularly the book of Genesis—regarding key issues such as creation, evolution, science, and the age of the earth" for those individuals seeking biblically-based alternatives to contemporary scientific accounts. As an illustration, ***Answers in Genesis*** offers us the following explanation regarding the age of the earth: "The earth is only a few thousand years old. That's a fact, plainly revealed in God's Word. So we should expect to find plenty of evidence for its youth. And that's what we find in the earth's geology, biology, paleontology, and even astronomy."[51]

Elsewhere this ministry comments on the infallibility of the Christian Bible: "Almost all theologians agree Scripture is in some measure God's revelation to the human race. But to allow that it contains error implies God has mishandled inspiration and has allowed His people to be deceived for centuries until modern scholars disentangled the confusion."[52]

Perhaps the best example of the rejection of scientific accounts by Fundamentalist Christians is the denial of Darwinian evolution, because it contradicts the literal description of the origins of humankind provided in Genesis. They further argue that data concerning the age of the earth and related matters obtained via radiometric dating, the fossil record and other means has been misinterpreted, is fraudulent or is not trustworthy for other reasons.[53]

[48]For a discussion outlining the differences between Christian Fundamentalists and Evangelical Christians, see https://www.whatchristianswanttoknow.com/what-is-the-difference-between-fundamentalists-and-evangelicals/.

[49]https://www.nae.net/evangelicals-anti-science/.

[50]http://www.nytimes.com/2011/10/18/opinion/the-evangelical-rejection-of-reason.html.

[51]https://answersingenesis.org/age-of-the-earth/.

[52]https://answersingenesis.org/is-the-bible-true/.

[53]https://answersingenesis.org/age-of-the-earth/dating-methods/.

That being said, many other religious traditions are far more open in their response to scientific discoveries and even the fundamentals of Transhumanism. Many Christian religions—Roman Catholicism is an example—accept evolution. In the 1950 encyclical Humani generis, Pope Pius XII indicated that there is no fundamental conflict between Christianity and the theory of evolution, but emphasized that the individual soul remained a direct divine creation and not the result of purely material events.[54]

Consider also the Mormon faith. The Church of Jesus Christ of Latter-Day Saints (LDS), also known as the Mormon Church, offers its believers the belief that humans can progress to become "like God". Their teaching on approaching godliness is explained at their official website[55]:

> Through the Atonement of Jesus Christ, all people may "progress toward perfection and ultimately realize their divine destiny."

and

> Lorenzo Snow, the Church's fifth President, coined a well-known couplet: "As man now is, God once was: As God now is, man may be."

To many Mormons, such theological teachings make transhumanism attractive. In fact, The Mormon Transhumanist Association is an organization that attempts to attempt to amalgamate (blend, fuse, syncretize) Mormonism and transhumanism. Information about the association may be obtained at http://transfigurism.org/ as well as at http://www.kurzweilai.net/man-will-become-like-god-say-mormons-and-transhumanists-in-salt-lake-city

Some other Christian groups also specifically embrace Transhumanist principles. For example, the Christian Transhumanist Association offers the following description of its core principles[56]:

> Christian Transhumanism takes the Christian idea that we are to become like Christ, and the Transhumanist idea that we should use science and technology in ethical ways to improve the world, and sees a strong connection.
>
> Christ exemplifies a life of service and love, and calls us to the work of healing the sick, feeding the hungry, and bringing life to the dead. Christ embodies acts of creativity and learning—the core of what we now call science and technology. Christ demonstrates that human nature is not limited by what we see around us—but is anchored in the continually unfolding image of God.
>
> Thus, to become like Christ involves using science and technology in ethical ways to improve the world.

Additionally, members of the Christian Transhumanist Association affirm the following specific beliefs and values[57]:

---

[54] Farrell, John (August 27, 2010). "Catholics and the Evolving Cosmos". The Wall Street Journal. https://www.wsj.com/articles/SB10001424052748703846604575447493644515142.

[55] https://www.lds.org/topics/becoming-like-god?lang=eng&old=true.

[56] https://www.christiantranshumanism.org/faq.

[57] https://www.christiantranshumanism.org/.

1. **We believe that God's mission involves the transformation and renewal of creation** including humanity, and that we are called by Christ to participate in that mission: working against illness, hunger, oppression, injustice, and death.
2. **We seek growth and progress along every dimension of our humanity:** spiritual, physical, emotional, mental—and at all levels: individual, community, society, world.
3. **We recognize science and technology as tangible expressions of our God-given impulse to explore and discover** and as a natural outgrowth of being created in the image of God.
4. **We are guided by Jesus' greatest commands** to "Love the Lord your God with all your heart, soul, mind, and strength...and love your neighbor as yourself."
5. **We believe that the intentional use of technology, coupled with following Christ, will empower us to become more human** across the scope of what it means to be creatures in the image of God.

8.18 Conclusion

The foregoing discussion strongly suggests that there exist clear neurological correlates to religious experience. While William James was perhaps the first to systematically address this issue, developments in neuropharmacology, brain imaging and epilepsy studies have led to a more complete mechanistic understanding of the nature of religious experience. Of course, still more needs to be done to improve our understanding of the neurological/psychological basis for religious experiences, especially with techniques such as f-MRI imaging. Regardless, it is reasonable to expect that atheists and believers will interpret any obtained insights through different philosophical lenses.

Chapter 9
Conclusions

> *A new scientific truth does not triumph by convincing its opponents and making them see the light, but rather because its opponents eventually die, and a new generation grows up that is familiar with it.*
>
> Max Planck (http://en.wikiquote.org/wiki/Max_Planck)

Abstract This chapter wraps up the discussion started in the first chapter of the book, on Questions such as "What does it mean to be human?", "When is a nonhuman a person?" and "When is a human not a person?", and reminds the reader that these questions have important moral implications for how we should treat higher animals, how we should care for humans with severe physical or cognitive impairments, and what forms of human enhancement are morally acceptable. The reader is reminded that two main schools of thought dominate this discourse: the transhumanists, who encourage human enhancement technologies and the bio-conservatives, who argue against artificial human enhancement. It is concluded that at its core, transhumanism is simply a vision of those who seek to ensure better lives for themselves and their children using new and existing biotechnologies and that the bio-conservative arguments against transhumanism are, for the most part, philosophically flawed.

9.1 Conclusions

The question "What does it mean to be human?" and the related questions "When is a nonhuman a person?" and "When is a human not a person?" have been the subject of philosophical deliberation for centuries. These questions have important moral implications for contemporary issues such as how we should treat higher animals, how we should care for humans with severe physical or cognitive impairments, and what forms of human enhancement are morally acceptable.

D. J. Doyle, *What Does it Mean to be Human? Life, Death, Personhood and the Transhumanist Movement*, Anticipation Science 3,
https://doi.org/10.1007/978-3-319-94950-5_9

Although recent scientific advances in fields such as genetic engineering, nanotechnology and biomedical electronics have added interesting new dimensions to these issues, the fact remains that the answers one obtains in response to such questions will depend substantially on the "lens" through which the discourse is examined.

Two main schools of thought dominate the discourse. The proponents of transhumanism would allow (indeed, even encourage) human enhancement technologies that would be expected to improve the human condition. They favor the voluntary deployment of safe and ethical enhancement technologies, regardless of whether or not any pathology needing correction might be present.

Opposing this school of thought are the bio-conservatives, who argue against artificial human enhancement on a variety of levels. These include safety concerns, concerns about fairness (as only the rich would have access to human enhancement technologies), concerns about these technologies contributing to the endangerment of human dignity, concerns about "slippery slope" issues, as well as a general feeling of repugnance for anything that is clearly not "natural."

As emphasized earlier, one fundamental issue in this debate centers on whether moral intuition should take precedence over reason and logic in making ethical decisions. Consider the example presented earlier: should the lives of severely deformed neonates ever be actively terminated? We saw that some moral philosophers such as Peter Singer regard this as sometimes acceptable, arguing that this follows as a consequence of utilitarian ethical principles. Other philosophers, we saw, argue that the natural repugnance of such an act should serve as a clear moral guide against such actions. However, as stressed earlier, any philosophical position emphasizing the role of moral intuition over logic and rationalism has potential flaws. From the time of the ancient Greeks philosophers have repeatedly emphasized the importance of logic in philosophical discourse. In Chap. 2, I similarly argue that reason and logic should take precedence over intuition in making ethical decisions. However, since I also accept the fact that reason and logic alone are not sufficient to establish a set of founding principles upon which to base ethical decision making, I can only conclude that the thorny issue cannot be definitively resolved by mere recourse to rational methods.

Chapter 3 introduced the transhumanist movement and related philosophical issues such as moral standing and personhood. The possibilities of genetic engineering as a means to fundamentally alter the nature of humankind were discussed, along with the concepts of animal-human hybrids, computer-human hybrids, and fully artificial beings. These ideas were presented to support the contention that there is no unique and unalterable biological form that constitutes humanity, and that our concept of what it means to be human will be forced to change in the face of new biotechnologies such as genetic engineering and computer-based neuroprostheses.

Another interesting question addressed in the book is whether or not precise pharmacological interventions might improve the human condition (see Chap. 4). As advances in the neurosciences develop we can expect that humankind might eventually come to enjoy the possibility of safely enjoying a "mood optimized" synthetic existence—an artificial paradise so to speak—made possible through advances in neuropharmacology. After exploring this possibility in some detail, I suggest that there is nevertheless the possibility that this might lead to an "inauthentic" existence that detracts from life's meaning (if life can be said to have a meaning), although I concede that this is in the end as much an empirical question as it is a philosophical one.

Chapter 5 addressed the surprisingly complex problem of establishing when a human can be said to be alive or said to be dead. Problems in identifying the exact point of death in both the traditional and neurological approaches to determining death were discussed, with philosophical difficulties in both approaches being identified.

Chapter 6 explored the concept of cryonic suspension as a philosophical problem in the context of establishing when a cryogenically suspended individual is still a live person and introduced the concept of "cryonic reincarnation" should brain substrate be restored without a person's life memories.

Chapter 7 examined the philosophical arguments raised against transhumanism and found them wanting on several grounds.

Chapter 8 examined transhumanism from the perspective of religious belief systems.

I would like to close with one final point. To those individuals who believe that the notions of human enhancement discussed herein are too far-fetched to merit serious discussion, let me provide them with a list of historical quotations that may help convince them otherwise (Fig. 9.1). It is almost inevitable that human enhancement technologies will eventually change humankind in very significant ways. Through developments in genetic engineering, computer-based neurotechnology, pharmacological discoveries and other methods, future humans will have opportunities for unprecedented life enhancements. At its core, transhumanism is simply a vision of those who seek to ensure better lives for themselves and their children using new and existing biotechnologies. I join them in that vision, but remain realistic about the obstacles to progress.

- No one will ever be able to measure nerve impulse speed. *Johannes Muller, German Physiologist, 1846*

- The abdomen, the chest and the brain will forever be shut from the intrusion of the wise and humane surgeon. *Sir John Eric Ericson, Surgeon to Queen Victoria, 1873*

- This `telephone' has too many shortcomings to be seriously considered as a practical form of communication. The device is inherently of no value to us. *Western Union internal memo, 1878*

- Everything that can be invented has been invented. *Charles H. Duell, Commissioner, U.S. Office of Patents, 1899*

- I must confess that my imagination, in spite even of spurring, refuses to see any sort of submarine doing anything but suffocating its crew and foundering at sea. *H. G. Wells, science fiction writer, 1901*

- Heavier-than-air flying machines are impossible. *Lord Kelvin (1824-1907), British mathematician and physicist*

- No flying machine will ever fly from New York to Paris. *Orville Wright, co-inventor of the airplane, 1908*

- While theoretically and technically television may be feasible, commercially and financially I consider it an impossibility, a development of which we need waste little time dreaming. *Lee DeForest, American radio pioneer, 1926.*

- There is no hope for the fanciful idea of reaching the Moon because of insurmountable barriers to escaping the Earth's gravity. *Dr. Forest Ray Moulton, University of Chicago astronomer, 1932*

Fig. 9.1 A sampling of historical quotations that suggest that famous inventors, scientists and engineers are sometimes poor at predicting the future of technology

- I think there is a world market for maybe five computers. *Thomas J. Watson Sr., IBM Chairman, 1943*

- To place a man in a multi-stage rocket and project him into the controlling gravitational field of the moon where the passengers can make scientific observations, perhaps land alive, and then return to earth--all that constitutes a wild dream worthy of Jules Verne. I am bold enough to say that such a man-made voyage will never occur regardless of all future advances. *Lee DeForest, American radio pioneer, 1957*

- There is no reason for any individual to have a computer in their home. *Ken Olson, President of Digital Corporation, 1977*

Fig. 9.1 (continued)

Appendix

A Short Transhumanist Bibliography

Transhumanist issues have been the subject of a considerable number of books; I have listed many of these in the following short bibliography. While the list is not exhaustive, it is highly representative of the transhumanist literature. The list is presented in order of publication date and includes both works aimed at the intelligent layman as well as works aimed at the professional academic philosopher. Note that this list is not offered as a kind of hermeneutic key or guide to the field but rather merely provided to acquaint the reader with some of the more readable accounts. In each case in this list I have tried to summarize the key message of the book in one or two sentences.

Gregory Stock. ***Redesigning Humans: Choosing our Genes, Changing our Future***. Mariner Books, 2003. *Inthis book Stock wants society to give our future children an edge, even if it involves profound changes of our biological makeup.*

Francis Fukuyama. ***Our Posthuman Future: Consequences of the Biotechnology Revolution***. Picador, 2003. *Inthis book the author identifies concerns that the foundations of liberal democracy may become eroded by unbridled biotechnology advances,and argues that humanbiotechnology must be regulated to prevent this and other undesirable outcomes.*

James Hughes. ***Citizen Cyborg: Why Democratic Societies Must Respond to the Redesigned Human of the Future***. Basic Books, 2004. *In this bookJamesHughes arguesthat while biomedical technologies likegenetic engineeringcan fundamentally improve our quality oflife, it is important that they be controlled democratically.*

Nicholas Agar. ***Liberal Eugenics: In Defence of Human Enhancement***. Wiley-Blackwell, 2004. *Inthisbook the author makes the case for ethical eugenic intervention through geneticenhancement of the humanorganism.*

D. J. Doyle, *What Does it Mean to be Human? Life, Death, Personhood and the Transhumanist Movement*, Anticipation Science3,
https://doi.org/10.1007/978-3-319-94950-5

Bill McKibben. ***Enough: Staying Human in an Engineered Age***. St. Martin's Griffin, 2004. *In this book McKibben argues that developments in genetic engineering, robotics, and nanotechnology threaten what it means to be human, and that for the sake of humanity we should critically consider the potential dehumanizing impact of such technologies before it is too late.*

Ronald Bailey. ***Liberation Biology: The Scientific and Moral Case for the Biotech Revolution***. Prometheus, 2005. *In this book the author takes a libertarian position, attacking both bioconservative thinkers such as Francis Fukuyama and Leon Kass, as well as critics of transhumanism from the political left.*

Simon Young. ***Designer Evolution: A Transhumanist Manifesto***. Prometheus Books, 2005. *In this book, Young argues for scientific rationality and a rejection of superstition and nihilism as ameans to encourage humanprogress.*

Ramez Naam. ***More Than Human: Embracing the Promise of Biological Enhancement***. Random House, 2005. *Written from a libertarian perspective, Naam reviews the state of the art ingenetic engineering, computer technology and other fields to make the case that biological enhancementwill improve the humancondition.*

Joel Garreau. ***Radical Evolution: The Promise and Peril of Enhancing Our Minds, Our Bodies—and What It Means to Be Human***. Broadway, 2006. *Like other books of this genre, the author seeks to present a balanced view. Along with a discussion of how humanity might be improved, Garreau identifies concerns raised by bioconservative thinkers such as FrancisFukuyama, Martin Rees, Bill McKibben, LeonKassand MichaelSandelthat while advances in genetics,nanotechnologyand robotics may help some segments of society, they also posegenuine threats to humansurvival itself.*

Ray Kurzweil. ***The Singularity is Near: When Humans Transcend Biology***. Penguin Books, 2006. *Kurzweil argues that humanity is at the threshold of an era ("the singularity") in which our biology will be combined with genetics,nanotechnologyand robotics to create anew form of humanorganism with almost no constraints on our mental, physical and moral capacity.*

Michael J. Sandel. ***The Case Against Perfection: Ethics in theAge of Genetic Engineering***. Belknap Press, 2009. *Inthis book the author argues that the pursuit of perfection via technological enhancement is morally problematic for reasons that go beyond arguments related to safety and social justice.Sandelcrafts an argument that the pursuit of perfection by such means represents a bid for mastery and control that ignores thegifted nature of humanpowers and achievements.*

Nicholas Agar. ***Humanity's End: Why We Should Reject Radical Enhancement***. Bradford, 2010. *In this book Agar argues that the radical interventions envisioned by some transhumanists could be far less positive than the rosy futures imagined by* *these thinkers, and that any transition to a posthuman future is fraught with moral challenges, such as the possibility that posthumans might rule tyrannically over ordinary humans.*

Allen Buchanan. *Better than Human: The Promise and Perils of Enhancing Ourselves*. Oxford University Press, 2011. *This book attempts to present a balanced view on the ethical dilemmas surrounding biomedical enhancement. In the book, the author argues that the debate over enhancement has been contaminated by false assumptions and disingenuous rhetoric, and that the existence of "design flaws" in our biological blueprint should be acknowledged. Consequently, humanity should be open to attempts to improve thedesign of the humanorganism (and with it,of course, humannature), while still recognizing that such attempts may on occasion fail calamitously.*

Gregory R. Hansell (Ed.). *Transhumanism and Its Critics*. Metanexus, 2011. *This volume is a collection of 16 essays presenting the cases for and against improving humanity via bioengineering methods.*

Julian Savulescu and Nick Bostrom (Eds.). *Human Enhancement*. Oxford University Press, 2011. *This anthology provides a comprehensive look atthe question of humanenhancement by artificial means, with commentaries by authorities from both transhumanist and bioconservative ('anti-meliorist') camps.*

Julian Savulescu and Ingmar Persson. *Unfit for the Future: The Need for Moral Enhancement*. Oxford University Press, 2012. *The authors argue that our moral shortcomings "are preventing our political institutions from acting effectively" and that artificial moral enhancement has become essential if humanity is to avoid catastrophe. Moral enhancement, the authors posit, "would enable us to act better for distant people, future generations, and non-human animals".*

Calvin Mercer and Tracy J. Trothen (Editors). *Religion and Transhumanism: The Unknown Future of Human Enhancement*. Praeger, 2014. *A collection of scholarly essays discussing the issue of whetherreligionandtranshumanismare at odds with each other.*

Daniel Berleant. *The Human Race to the Future: What Could Happen—and What to Do*. Lifeboat Foundation, 2014. *The author offers us visions of possible futures ranging from the near-present to many millions of years from now.*

David Livingstone. *Transhumanism: The History of a Dangerous Idea*. CreateSpace Independent Publishing Platform, 2015. *Argues that "transhumanism is an occult project, rooted in Rosicrucianism and Freemasonry, and derived from the Kabbalah, which asserts that humanity is evolving intellectually, towards a point in time when man will become God."*

Juan Enriquez. *Evolving Ourselves: Redesigning the Future of Humanity—One Gene at a Time Revised Edition*. Current, 2016. *Explains how developments in genetic engineering such as CRISPR will allow humanity to shape its evolution.*

James Kozubek. ***Modern Prometheus: Editing the Human Genome with Crispr-Cas9***. Cambridge University Press, 2016. *Another book on how CRISPR will allow humanity to shape its evolution.*

Andrew Pilsch. ***Transhumanism: Evolutionary Futurism and the Human Technologies of Utopia***. Univ of Minnesota Press, 2017. *Presents a generally positive view of the future possibilities offered by the Transhumanist movement.*

References

Adam, Yehuda G. 2007. Justice in Nuremberg: The Doctors' Trial—60 Years Later. A Reminder. *The Israel Medical Association Journal* 9 (3): 194–195.

Adamenko, N.P. 1969. Technic of Experimental Resuscitation of Dogs with a Variant of the Method of Extracorporeal Circulation Using S. S. Briukhonenko's Autojector Pump and a Live Parabiotic Donor. *Patologicheskaia Fiziologiia I Eksperimental'naia Terapiia* 13 (3): 69–71.

Adelman, Barry Eshkol. 2007. An Underdiscussed Aspect of Chomsky (1959). *The Analysis of Verbal Behavior* 23: 29–34. https://doi.org/10.1007/BF03393044.

Ad Hoc Committee of the Harvard Medical School to Examine the Definition of Brain Death. 1968. A Definition of Irreversible Coma. Report of the Ad Hoc Committee of the Harvard Medical School to Examine the Definition of Brain Death. *The Journal of the American Medical Association* 205 (6): 337–340.

Agar, N. 2004. *Liberal Eugenics. In Defence of Human Enhancement*. Hoboken, NJ: Blackwell.

Aichner, F., C. Adelwöhrer, and H.P. Haring. 2002. Rehabilitation Approaches to Stroke. *Journal of Neural Transmission. Supplementum* (63): 59–73.

Alexander, J.B. 2001. An Overview of the Future of Non-lethal Weapons. *Medicine, Conflict, and Survival* 17 (3): 180–193. https://doi.org/10.1080/13623690108409578.

Allmark, Peter. 2006. An Argument for the Use of Aristotelian Method in Bioethics. *Medicine, Health Care, and Philosophy* 9 (1): 69–79. https://doi.org/10.1007/s11019-005-7225-x.

Angell, M. 1990. The Nazi Hypothermia Experiments and Unethical Research Today. *The New England Journal of Medicine* 322 (20): 1462–1464. https://doi.org/10.1056/NEJM199005173222011.

Archer, J. 1991. The Influence of Testosterone on Human Aggression. *British Journal of Psychology* 82 (Pt 1) (February): 1–28.

Archer, John. 2006. Testosterone and Human Aggression: An Evaluation of the Challenge Hypothesis. *Neuroscience and Biobehavioral Reviews* 30 (3): 319–345. https://doi.org/10.1016/j.neubiorev.2004.12.007.

Arries, E. 2005. Virtue Ethics: An Approach to Moral Dilemmas in Nursing. *Curationis* 28 (3): 64–72.

Arya, Neil. 2007. Is Military Action Ever Justified? A Physician Defends the 'Responsibility to Protect'. *Medicine, Conflict, and Survival* 23 (3): 172–188. https://doi.org/10.1080/13623690701417311.

Asimov, I. 1991. *I Robot*. New York: Spectra.

Atkins, Kim. 2005. *Self and Subjectivity*. Malden, MA: Blackwell Publishing.

Atran, Scott, and Jeremy Ginges. 2012. Religious and Sacred Imperatives in Human Conflict. *Science* 336 (6083): 855–857. https://doi.org/10.1126/science.1216902.

D. J. Doyle, *What Does it Mean to be Human? Life, Death, Personhood and the Transhumanist Movement*, Anticipation Science3,
https://doi.org/10.1007/978-3-319-94950-5

Baier, K. 1958. *The Moral Point of View*. NY: Cornell University Press.

Baldry, C., S.B. Backman, P. Metrakos, J. Tchervenkov, J. Barkun, and A. Moore. 2000. Liver Transplantation in a Jehovah's Witness with Ankylosing Spondylitis. *Canadian Journal of Anaesthesia* 47 (7): 642–646.

Baron, Leonard, Sam D. Shemie, Jeannie Teitelbaum, and Christopher James Doig. 2006. Brief Review: History, Concept and Controversies in the Neurological Determination of Death. *Canadian Journal of Anaesthesia* 53 (6): 602–608. https://doi.org/10.1007/BF03021852.

Barondess, J.A. 1996. Medicine Against Society. Lessons from the Third Reich. *The Journal of the American Medical Association* 276 (20): 1657–1661.

Bartels, A., and S. Zeki. 2000. The Neural Basis of Romantic Love. *Neuroreport* 11 (17): 3829–3834. https://doi.org/10.1097/00001756-200011270-00046.

Bartels, Andreas, and Semir Zeki. 2004. The Neural Correlates of Maternal and Romantic Love. *Neuroimage* 21 (3): 1155–1166. https://doi.org/10.1016/j.neuroimage.2003.11.003.

Baschetti, Riccardo. 2005. Evolutionary, Biological Origins of Morality: Implications for Research with Human Embryonic Stem Cells. *Stem Cells and Development* 14 (3): 239–247. https://doi.org/10.1089/scd.2005.14.239.

Baumard, Nicolas, and Pascal Boyer. 2013. Explaining Moral Religions. *Trends in Cognitive Sciences* 17 (6): 272–280. https://doi.org/10.1016/j.tics.2013.04.003.

Beauchamp, Tom. 2004. The Legacy and the Future. 30 Years after the Belmont Report, Beauchamp Sets the Record Straight. *Protecting Human Subjects* (10): 1–3.

Beauchamp, Tom L. 1995. Principlism and Its Alleged Competitors. *Kennedy Institute of Ethics Journal* 5 (3): 181–198.

Beauchamp, Tom L. 1999. The Failure of Theories of Personhood. *Kennedy Institute of Ethics Journal* 9 (4): 309–324.

Beauchamp, Tom L., and James F. Childress. 2013. *Principles of Biomedical Ethics*, 7th ed. New York: Oxford University Press.

Beauregard, Mario, and Vincent Paquette. 2008. EEG Activity in Carmelite Nuns During a Mystical Experience. *Neuroscience Letters* 444 (1): 1–4. https://doi.org/10.1016/j.neulet.2008.08.028.

Becker, Gary J. 2005. Human Subjects Investigation: Timeless Lessons of Nuremberg and Tuskegee. *Journal of the American College of Radiology* 2 (3): 215–217. https://doi.org/10.1016/j.jacr.2004.11.018.

Begley, Ann M. 2008. Truth-telling, Honesty and Compassion: A Virtue-based Exploration of a Dilemma in Practice. *International Journal of Nursing Practice* 14 (5): 336–341. https://doi.org/10.1111/j.1440-172X.2008.00706.x.

Begley, Ann Marie. 2005. Practising Virtue: A Challenge to the View that a Virtue Centred Approach to Ethics Lacks Practical Content. *Nursing Ethics* 12 (6): 622–637. https://doi.org/10.1191/0969733005ne832oa.

Beit-Hallahmi, B., and M. Argyle. 1977. Religious Ideas and Psychiatric Disorders. *The International Journal of Social Psychiatry* 23 (1): 26–30. https://doi.org/10.1177/002076407702300105.

Bencherif, Merouane, and Patrick M. Lippiello. 2010. Alpha7 Neuronal Nicotinic Receptors: The Missing Link to Understanding Alzheimer's Etiopathology? *Medical Hypotheses* 74 (2): 281–285. https://doi.org/10.1016/j.mehy.2009.09.011.

Benson, D.F. 1991. The Geschwind Syndrome. *Advances in Neurology* 55: 411–421.

Benson, K.T. 1989. The Jehovah's Witness Patient: Considerations for the Anesthesiologist. *Anesthesia and Analgesia* 69 (5): 647–656.

Berg, Paul. 2008. Meetings That Changed the World: Asilomar 1975: DNA Modification Secured. *Nature* 455 (7211): 290–291. https://doi.org/10.1038/455290a.

Berger, R.L. 1990. Nazi Science—The Dachau Hypothermia Experiments. *The New England Journal of Medicine* 322 (20): 1435–1440. https://doi.org/10.1056/NEJM199005173222006.

Bernat, James L. 2005. Medical Futility: Definition, Determination, and Disputes in Critical Care. *Neurocritical Care* 2 (2): 198–205. https://doi.org/10.1385/NCC:2:2:198.

Bernat, J.L., A.M. D'Alessandro, F.K. Port, T.P. Bleck, S.O. Heard, J. Medina, S.H. Rosenbaum, et al. 2006. Report of a National Conference on Donation after Cardiac Death. *American Journal of Transplantation* 6 (2): 281–291. https://doi.org/10.1111/j.1600-6143.2005.01194.x.

Bernstein, Mark, and Barry Brown. 2004. Doctors' Duty to Disclose Error: a Deontological or Kantian Ethical Analysis. *The Canadian Journal of Neurological Sciences. Le Journal Canadien Des Sciences Neurologiques* 31 (2): 169–174.

Best, Benjamin P. 2009. Nuclear DNA Damage as a Direct Cause of Aging. *Rejuvenation Research* 12 (3): 199–208. https://doi.org/10.1089/rej.2009.0847.

Bewernick, Bettina H., René Hurlemann, Andreas Matusch, Sarah Kayser, Christiane Grubert, Barbara Hadrysiewicz, Nikolai Axmacher, et al. 2010. Nucleus Accumbens Deep Brain Stimulation Decreases Ratings of Depression and Anxiety in Treatment-resistant Depression. *Biological Psychiatry* 67 (2): 110–116. https://doi.org/10.1016/j.biopsych.2009.09.013.

Bhavsar, Vishal, and Dinesh Bhugra. 2008. Religious Delusions: Finding Meanings in Psychosis. *Psychopathology* 41 (3): 165–172. https://doi.org/10.1159/000115954.

Black, E. 2003. *War Against the Weak. Eugenics and America's Campaign to Create a Master Race*. NY: Four Walls Eight Windows.

Blackburn, Elizabeth, and Janet Rowley. 2004. Reason as Our Guide. *PLoS Biology* 2 (4): E116. https://doi.org/10.1371/journal.pbio.0020116.

Bliss, T.V., and G.L. Collingridge. 1993. A Synaptic Model of Memory: Long-term Potentiation in the Hippocampus. *Nature* 361 (6407): 31–39. https://doi.org/10.1038/361031a0.

Bloom, Paul. 2012. Religion, Morality, Evolution. *Annual Review of Psychology* 63: 179–199. https://doi.org/10.1146/annurev-psych-120710-100334.

Blum, Kenneth, Marlene Oscar-Berman, Abdalla Bowirrat, John Giordano, Margaret Madigan, Eric R. Braverman, Debmayla Barh, Mary Hauser, Joan Borsten, and Thomas Simpatico. 2012. Neuropsychiatric Genetics of Happiness, Friendships, and Politics: Hypothesizing Homophily ('Birds of a Feather Flock Together') as a Function of Reward Gene Polymorphisms. *Journal of Genetic Syndromes & Gene Therapy* 3 (112). https://doi.org/10.4172/2157-7412.1000112.

Blumenthal, Ralph, and Judith Miller. 1999. Japanese Germ-war Atrocities: A Half-century of Stonewalling the World. *The New York Times*, March, A12.

Bolsin, Stephen. 2002. Mortality and Volume of Cases in Paediatric Cardiac Surgery. Paper Confirms Poor Quality of Paediatric Heart Surgery at Bristol During 1991–5. *BMJ (Clinical Research Ed.)* 324 (7345): 1095–1096.

Bolt, L.L.E. 2007. True to Oneself? Broad and Narrow Ideas on Authenticity in the Enhancement Debate. *Theoretical Medicine and Bioethics* 28 (4): 285–300. https://doi.org/10.1007/s11017-007-9039-8.

Borgens, R.B., and R. Shi. 2000. Immediate Recovery from Spinal Cord Injury through Molecular Repair of Nerve Membranes with Polyethylene Glycol. *The FASEB Journal* 14 (1): 27–35.

Bostrom, N. 2002. Analyzing Human Extinction Scenarios and Related Hazards. *Journal of Evolution and Technology* 9 (1): 1–37.

Bostrom, Nick. 2003. Human Genetic Enhancements: A Transhumanist Perspective. *The Journal of Value Inquiry* 37 (4): 493–506.

Bostrom, Nick. 2005. In Defense of Posthuman Dignity. *Bioethics* 19 (3): 202–214.

Bowlby, J. 1969. *Attachment and Loss Vol. 1*. 2 vols. Attachment. London: Basic Book.

Bowlby, J. 1976. *Attachment and Loss Vol. II*. Separation: Anxiety and Anger.

Bowlby, J. 1988. *A Secure Base: Clinical Applications of Attachment Theory*. London: Psychology Press.

Boyer, Pascal. 2003. Religious Thought and Behaviour as By-products of Brain Function. *Trends in Cognitive Sciences* 7 (3): 119–124. https://doi.org/10.1016/S1364-6613(03)00031-7.

Brassington, Iain. 2007. John Harris' Argument for a Duty to Research. *Bioethics* 21 (3): 160–168. https://doi.org/10.1111/j.1467-8519.2007.00539.x.

Bretherton, Inge. 1992. The Origins of Attachment Theory: John Bowlby and Mary Ainsworth. *Developmental Psychology* 28 (5): 759–775. https://doi.org/10.1037/0012-1649.28.5.759.

Bridger, S., K. Henderson, E. Glucksman, A.J. Ellis, J.A. Henry, and R. Williams. 1998. Deaths from Low Dose Paracetamol Poisoning. *BMJ (Clinical Research Ed.)* 316 (7146): 1724–1725.

Brodtkorb, Eylert, and Karl Otto Nakken. 2015. The Relationship Between Epilepsy and Religiosity Illustrated by the Story of the Visionary Mystic Wise-Knut. *Epilepsy & Behavior* 46 (May): 99–102. https://doi.org/10.1016/j.yebeh.2015.04.017.

Brody, Howard, and Arlene Macdonald. 2013. Religion and Bioethics: Toward an Expanded Understanding. *Theoretical Medicine and Bioethics* 34 (2): 133–145. https://doi.org/10.1007/s11017-013-9244-6.

Brorson, J.R., and K. Brewer. 1988. St. Paul and Temporal Lobe Epilepsy. *Journal of Neurology, Neurosurgery, and Psychiatry* 51 (6): 886–887.

Brukhonenko, S, and S. Tchetchuline. 1929a. Expériences Avec La Têteisolée Du Chien. I. Technique et Conditions Des Experiénces. *Journal de physiologie et de pathologie générale* 29 (1): 31–45.

Brukhonenko, S, and S. Tchetchuline. 1929b. Expériences Avec La Têteisolée Du Chien. II. Résultats Des Experiénces. *Journal de physiologie et de pathologie générale* 29 (1): 64–79.

Buckley, P. 1981. Mystical Experience and Schizophrenia. *Schizophrenia Bulletin* 7 (3): 516–521.

Bulkeley, Kelly. 2016. *Big Dreams: The Science of Dreaming and the Origins of Religion*. Oxford, UK: Oxford University Press.

Burns, Jeffrey P., and Robert D. Truog. 2007. Futility: A Concept in Evolution. *Chest* 132 (6): 1987–1993. https://doi.org/10.1378/chest.07-1441.

Burrows, Richard, R. David Crippen, Phillip Dellinger, David F. Kelly, Stephen Streat, and Leslie M. Whetstine. 2002. Ethics Roundtable: Using New, Expensive Drugs. *Critical Care* 6 (6): 473–478.

Burt, Robert A. 2002. The Medical Futility Debate: Patient Choice, Physician Obligation, and End-of-life Care. *Journal of Palliative Medicine* 5 (2): 249–254. https://doi.org/10.1089/109662102753641223.

Buxton, Jane A., and Naomi A. Dove. 2008. The Burden and Management of Crystal Meth Use. *Canadian Medical Association Journal* 178 (12): 1537–1539. https://doi.org/10.1503/cmaj.071234.

Cakic, V. 2009. Smart Drugs for Cognitive Enhancement: Ethical and Pragmatic Considerations in the Era of Cosmetic Neurology. *Journal of Medical Ethics* 35 (10): 611–615. https://doi.org/10.1136/jme.2009.030882.

Caldwell, John A., J. Lynn Caldwell, and Kecia K. Darlington. 2003. Utility of Dextroamphetamine for Attenuating the Impact of Sleep Deprivation in Pilots. *Aviation, Space, and Environmental Medicine* 74 (11): 1125–1134.

California. Court of Appeal, Second District, Division 6. 1992. Donaldson V. van de Kamp. *West's California Reporter* 4 (January): 59–65.

Canavero, Sergio. 2015. The 'Gemini' Spinal Cord Fusion Protocol: Reloaded. *Surgical Neurology International* 6 (February): 18. https://doi.org/10.4103/2152-7806.150674.

Canavero, Sergio, and Xiaoping Ren. 2016. Houston, GEMINI Has Landed: Spinal Cord Fusion Achieved. *Surgical Neurology International* 7 (Suppl 24): S626–S628. https://doi.org/10.4103/2152-7806.190473.

Capps, D. 2001. *Freud and Freudians on Religion: A Reader*. New Haven: Yale University Press.

Carhart-Harris, Robin L., Suresh Muthukumaraswamy, Leor Roseman, Mendel Kaelen, Wouter Droog, Kevin Murphy, Enzo Tagliazucchi, et al. 2016. Neural Correlates of the LSD Experience Revealed by Multimodal Neuroimaging. *Proceedings of the National Academy of Sciences of the United States of America* 113 (17): 4853–4858. https://doi.org/10.1073/pnas.1518377113.

Carnevale, Franco A. 2009. A Conceptual and Moral Analysis of Suffering. *Nursing Ethics* 16 (2): 173–183. https://doi.org/10.1177/0969733008100076.

Carter, C.S., and M. Altemus. 1997. Integrative Functions of Lactational Hormones in Social Behavior and Stress Management. *Annals of the New York Academy of Sciences* 807 (January): 164–174.

Cassell, E.J. 2000. The Principles of the Belmont Report Revisited. How Have Respect for Persons, Beneficence, and Justice Been Applied to Clinical Medicine? *The Hastings Center Report* 30 (4): 12–21.

Cavalieri, P., and P. Singer. 1994. *The Great Ape Project: Equality Beyond Humanity*. New York: St. Martin's Griffin.

Chalmers, Don, and Robert Schwartz. 1993. Rogers V. Whitaker and Informed Consent in Australia: A Fair Dinkum Duty of Disclosure. *Medical Law Review* 1 (2): 139–159.

Chan, Sarah, and John Harris. 2010. Consequentialism Without Consequences: Ethics and Embryo Research. *Cambridge Quarterly of Healthcare Ethics : CQ : The International Journal of Healthcare Ethics Committees* 19 (1): 61–74. https://doi.org/10.1017/S0963180109990259.

Chan, Sarah, and John Harris. 2011. Moral Enhancement and Pro-social Behaviour. *Journal of Medical Ethics* 37 (3): 130–131. https://doi.org/10.1136/jme.2010.041434.

Charlton, Bruce G. 2008. Genospirituality: Genetic Engineering for Spiritual and Religious Enhancement. *Medical Hypotheses* 71 (6): 825–828. https://doi.org/10.1016/j.mehy.2008.06.006.

Chatterjee, Anjan. 2004. Cosmetic Neurology: The Controversy over Enhancing Movement, Mentation, and Mood. *Neurology* 63 (6): 968–974.

Chen, Y.F. 1997. Japanese Death Factories and the American Cover-up. *Cambridge Quarterly of Healthcare Ethics : CQ : The International Journal of Healthcare Ethics Committees* 6 (2): 240–242.

Chi, Richard P., Felipe Fregni, and Allan W. Snyder. 2010. Visual Memory Improved by Non-invasive Brain Stimulation. *Brain Research* 1353 (September): 168–175. https://doi.org/10.1016/j.brainres.2010.07.062.

Cho, M.K., D. Magnus, A.L. Caplan, and D. McGee. 1999. Policy Forum: Genetics. Ethical Considerations in Synthesizing a Minimal Genome. *Science* 286 (5447): 2087.

Clark, R.A. 1981. Self-mutilation Accompanying Religious Delusions: A Case Report and Review. *The Journal of Clinical Psychiatry* 42 (6): 243–245.

Cleland, Carol E., and Christopher F. Chyba. 2002. Defining 'Life'. *Origins of Life and Evolution of the Biosphere* 32 (4): 387–393.

Clemens, Kelly J., Jennifer L. Cornish, Kong M. Li, Glenn E. Hunt, and Iain S. McGregor. 2005. MDMA ('Ecstasy') and Methamphetamine Combined: Order of Administration Influences Hyperthermic and Long-term Adverse Effects in Female Rats. *Neuropharmacology* 49 (2): 195–207. https://doi.org/10.1016/j.neuropharm.2005.03.002.

Clemens, Kelly J., Petra S. Van Nieuwenhuyzen, Kong M. Li, Jennifer L. Cornish, Glenn E. Hunt, and Iain S. McGregor. 2004. MDMA ('ecstasy'), Methamphetamine and Their Combination: Long-term Changes in Social Interaction and Neurochemistry in the Rat. *Psychopharmacology* 173 (3–4): 318–325. https://doi.org/10.1007/s00213-004-1786-x.

Clouser, K.D., and B. Gert. 1990. A Critique of Principlism. *The Journal of Medicine and Philosophy* 15 (2): 219–236.

Clouser, K. Danner. 1995. Common Morality as an Alternative to Principlism. *Kennedy Institute of Ethics Journal* 5 (3): 219–236.

Conner, James M., Kevin M. Franks, Andrea K. Titterness, Kyle Russell, David A. Merrill, Brian R. Christie, Terrence J. Sejnowski, and Mark H. Tuszynski. 2009. NGF Is Essential for Hippocampal Plasticity and Learning. *The Journal of Neuroscience* 29 (35): 10883–10889. https://doi.org/10.1523/JNEUROSCI.2594-09.2009.

Corlett, J.A. 2001. Is There a Moral Duty to Die? *Health Care Analysis: HCA: Journal of Health Philosophy and Policy* 9 (1): 41–63. https://doi.org/10.1023/A:1011355028021.

Costanzo, Jon P., and Richard E. Lee. 2005. Cryoprotection by Urea in a Terrestrially Hibernating Frog. *The Journal of Experimental Biology* 208 (Pt 21): 4079–4089. https://doi.org/10.1242/jeb.01859.

Costanzo, Jon P., and Richard E. Lee. 2008. Urea Loading Enhances Freezing Survival and Postfreeze Recovery in a Terrestrially Hibernating Frog. *The Journal of Experimental Biology* 211 (Pt 18): 2969–2975. https://doi.org/10.1242/jeb.019695.

Côté, A. 2000. Telling the Truth? Disclosure, Therapeutic Privilege and Intersexuality in Children. *Health Law Journal* 8: 199–216.

Cox, Ashley, and Robert Zhong. 2005. Current Advances in Xenotransplantation. *Hepatobiliary & Pancreatic Diseases International* 4 (4): 490–494.

D'Aquili, Eugene, and Andrew Newberg. 1999. *The Mystical Mind: Probing the Biology of Religious Experience*. Minneapolis, MN: Fortress Press.

Das, Saibal, Preeti Barnwal, Anand Ramasamy, Sumalya Sen, and Somnath Mondal. 2016. Lysergic Acid Diethylamide: A Drug of 'Use'? *Therapeutic Advances in Psychopharmacology* 6 (3): 214–228. https://doi.org/10.1177/2045125316640440.

Davey, L.M. 2001. The Oath of Hippocrates: An Historical Review. *Neurosurgery* 49 (3): 554–566.

De Boer, Sietse F., Doretta Caramaschi, Deepa Natarajan, and Jaap M. Koolhaas. 2009. The Vicious Cycle Towards Violence: Focus on the Negative Feedback Mechanisms of Brain Serotonin Neurotransmission. *Frontiers in Behavioral Neuroscience* 3 (November): 52. https://doi.org/10.3389/neuro.08.052.2009.

De Grey, A.D. 1997. A Proposed Refinement of the Mitochondrial Free Radical Theory of Aging. *Bioessays: News and Reviews in Molecular, Cellular and Developmental Biology* 19 (2): 161–166. https://doi.org/10.1002/bies.950190211.

Debiec, Jacek. 2007. From Affiliative Behaviors to Romantic Feelings: A Role of Nanopeptides. *FEBS Letters* 581 (14): 2580–2586. https://doi.org/10.1016/j.febslet.2007.03.095.

Deckel, A.W. 1996. Behavioral Changes in Anolis Carolinensis Following Injection with Fluoxetine. *Behavioural Brain Research* 78 (2): 175–182.

Dees, Richard H. 2004. Slippery Slopes, Wonder Drugs, and Cosmetic Neurology: The Neuroethics of Enhancement. *Neurology* 63 (6): 951–952.

DeGrazia, David. 2005. Enhancement Technologies and Human Identity. *The Journal of Medicine and Philosophy* 30 (3): 261–283. https://doi.org/10.1080/03605310590960166.

De las Casas, Bartolomé. 1992. *The Devastation of the Indies: A Brief Account*. Baltimore: Johns Hopkins University Press.

Delgado, J.M.R. 1969. *Physical Control of the Mind: Toward a Psychocivilized Society*. NY: Harper & Row.

DeMarco, D. 2003. Peter Singer: Architect of the Culture of Death. *Social Justice Review* 94: 154–157.

Derr, J.S., and M.A. Persinger. 1989. Geophysical Variables and Behavior: LIV. Zeitoun (Egypt) Apparitions of the Virgin Mary as Tectonic Strain-induced Luminosities. *Perceptual and Motor Skills* 68 (1): 123–128. https://doi.org/10.2466/pms.1989.68.1.123.

Devinsky, Orrin, and George Lai. 2008. Spirituality and Religion in Epilepsy. *Epilepsy & Behavior* 12 (4): 636–643. https://doi.org/10.1016/j.yebeh.2007.11.011.

Diamant, J.C. 2002. The Revised Declaration of Helsinki—Is Justice Served? *International Journal of Clinical Pharmacology and Therapeutics* 40 (2): 76–83.

Dickman, Steven. 1988. US Role Alleged in Cover-up of Researchers Guilty of War Crimes. *Nature* 335 (6190): 481. https://doi.org/10.1038/335481a0.

Dodman, N.H., R. Donnelly, L. Shuster, P. Mertens, W. Rand, and K. Miczek. 1996. Use of Fluoxetine to Treat Dominance Aggression in Dogs. *Journal of the American Veterinary Medical Association* 209 (9): 1585–1587.

Doebeli, Michael, and Iaroslav Ispolatov. 2010. A Model for the Evolutionary Diversification of Religions. *Journal of Theoretical Biology* 267 (4): 676–684. https://doi.org/10.1016/j.jtbi.2010.09.013.

Doig, Christopher James, and Graeme Rocker. 2003. Retrieving Organs from Non-heart-beating Organ Donors: A Review of Medical and Ethical Issues. *Canadian Journal of Anaesthesia* 50 (10): 1069–1076. https://doi.org/10.1007/BF03018376.

Donaldson, Zoe R., and Larry J. Young. 2008. Oxytocin, Vasopressin, and the Neurogenetics of Sociality. *Science* 322 (5903): 900–904. https://doi.org/10.1126/science.1158668.

Doyle, D. 2007. Ergonomics, Patient Safety, and Engineering Ethics: A Case Study and Cautionary Tale. *Journal of Long-term Effects of Medical Implants* 17 (1): 27–33.
Doyle, D.J., and A. Robichaud. 2004. Inconsistencies in the Ethical Declaration of Death. *Canadian Journal of Anaesthesia* 51 (3): 280.
Drexler, E. 1986. *The Engines of Creation. The Coming Age of Nanotechnology*. NY: Doubleday.
Dreyfus, H. 1979. *What Computers Can't Do*. Cambridge, MA: MIT Press.
Dreyfus, H. 1986. *Mind over Machine: The Power of Human Intuition and Expertise in the Era of the Computer*. Oxford, UK: Blackwell.
Dreyfus, H. 1992. *What Computers Still Can't Do*. Cambridge, MA: MIT Press.
Drucker, E. 1999. Drug Prohibition and Public Health: 25 Years of Evidence. *Public Health Reports (Washington, D.C. : 1974)* 114 (1): 14–29.
Dudai, Yadin. 2004. The Neurobiology of Consolidations, or, How Stable Is the Engram? *Annual Review of Psychology* 55: 51–86. https://doi.org/10.1146/annurev.psych.55.090902.142050.
Egonsson, Dan. 2009. Death and Irreversibility. *Reviews in the Neurosciences* 20 (3–4): 275–281.
Elder, L. 2000. Why Some Jehovah's Witnesses Accept Blood and Conscientiously Reject Official Watchtower Society Blood Policy. *Journal of Medical Ethics* 26 (5): 375–380.
Elliott, C. 1998. The Tyranny of Happiness: Ethics and Cosmetic Psychopharmacology. In *Enhancing Human Traits: Ethical and Social Implications*, ed. E. Parens, 177–188. Washington, D.C.: Georgetown University Press.
Elliott, C. 1998. The Tyranny of Happiness: Ethics and Cosmetic Psychopharmacology. In *Enhancing Human Traits: Ethical and Social Implications*, ed. E. Parens, 177–188. Washington, DC: Georgetown University Press.
Elliott, C. 2003. *Better Than Well. American Medicine Meets the American Dream*. NY: W.W. Norton & Company.
Elliott, Richard. 2004. Bangkok 2004. Drug Control, Human Rights, and Harm Reduction in the Age of AIDS. *HIV/AIDS Policy & Law Review* 9 (3): 86–90.
Ellis, Mark R. 2002. Challenges Posed by a Scientific Approach to Spiritual Issues. *The Journal of Family Practice* 51 (3): 259–260.
Elwyn, Todd S., Michael D. Fetters, Hiroki Sasaki, and Tsukasa Tsuda. 2002. Responsibility and Cancer Disclosure in Japan. *Social Science & Medicine* 54 (2): 281–293.
Epstein, Miran. 2007. The Ethics of Poverty and the Poverty of Ethics: The Case of Palestinian Prisoners in Israel Seeking to Sell Their Kidneys in Order to Feed Their Children. *Journal of Medical Ethics* 33 (8): 473–474. https://doi.org/10.1136/jme.2006.017855.
Erin, Charles A., and John Harris. 2003. An Ethical Market in Human Organs. *Journal of Medical Ethics* 29 (3): 137–138.
Erler, Alexandre. 2011. Does Memory Modification Threaten Our Authenticity? *Neuroethics* 4 (3): 235–249. https://doi.org/10.1007/s12152-010-9090-4.
Etchells, E., G. Sharpe, M.M. Burgess, and P.A. Singer. 1996. Bioethics for Clinicians: 2. Disclosure. *Canadian Medical Association Journal* 155 (4): 387–391.
Ettinger, R. 2005. *The Prospect of Immortality*. Palo Alto, CA: Ria University Press.
Ettinger, R. 2005. *Man into Superman: The Startling Potential of Human Evolution – And How to Be Part of It*. Palo Alto, CA: Ria University Press.
Evans, Martin. 2005. Ethical Sourcing of Human Embryonic Stem Cells—Rational Solutions? *Nature Reviews. Molecular Cell Biology* 6 (8): 663–667. https://doi.org/10.1038/nrm1698.
Fahy, Gregory M., and Brian Wowk. 2015. Principles of Cryopreservation by Vitrification. *Methods in Molecular Biology* 1257: 21–82. https://doi.org/10.1007/978-1-4939-2193-5_2.
Fahy, Gregory M., Brian Wowk, Roberto Pagotan, Alice Chang, John Phan, Bruce Thomson, and Laura Phan. 2009. Physical and Biological Aspects of Renal Vitrification. *Organogenesis* 5 (3): 167–175.
Farr, A.D. 1980. Early Opposition to Obstetric Anaesthesia. *Anaesthesia* 35 (9): 896–907.
Fehrenbacher, D.E. 1981. *Slavery, Law, and Politics: The Dred Scott Case in Historical Perspective*. New York: Oxford University Press.

Ferris, C.F. 1996. Serotonin Diminishes Aggression by Suppressing the Activity of the Vasopressin System. *Annals of the New York Academy of Sciences* 794 (September): 98–103.

Fiore, Arlene M., Vaishali Naik, Dominick V. Spracklen, Allison Steiner, Nadine Unger, Michael Prather, Dan Bergmann, et al. 2012. Global Air Quality and Climate. *Chemical Society Reviews* 41 (19): 6663–6683. https://doi.org/10.1039/c2cs35095e.

Forsberg, Joanna Stjernschantz, Mats G. Hansson, and Stefan Eriksson. 2014. Why Participating in (certain) Scientific Research Is a Moral Duty. *Journal of Medical Ethics* 40 (5): 325–328. https://doi.org/10.1136/medethics-2012-100859.

Frankfurt, H.G. 1971. Freedom of the Will and the Concept of a Person. *The Journal of Philosophy* 68: 5–7.

Frati, Paola, Chrystalla Kyriakou, Alessandro Del Rio, Enrico Marinelli, Gianluca Montanari Vergallo, Simona Zaami, and Francesco P. Busardò. 2015. Smart Drugs and Synthetic Androgens for Cognitive and Physical Enhancement: Revolving Doors of Cosmetic Neurology. *Current Neuropharmacology* 13 (1): 5–11. https://doi.org/10.2174/1570159X13666141210221750.

French, R.M. 2000. The Turing Test: The First 50 Years. *Trends in Cognitive Sciences* 4 (3): 115–122.

Friedlander, W.J. 1982. Oaths Given by U.S. and Canadian Medical Schools, 1977: Profession of Medical Values. *Social Science & Medicine* 16 (1): 115–120.

Friedlander, Walter J. 1995. The Evolution of Informed Consent in American Medicine. *Perspectives in Biology and Medicine* 38 (3): 498–510.

Friedman, David, Bracha Rager-Zisman, Eitan Bibi, and Alex Keynan. 2010. The Bioterrorism Threat and Dual-use Biotechnological Research: An Israeli Perspective. *Science and Engineering Ethics* 16 (1): 85–97. https://doi.org/10.1007/s11948-008-9075-4.

Friedman, E.A., and A.L. Friedman. 2006. Payment for Donor Kidneys: Pros and Cons. *Kidney International* 69 (6): 960–962. https://doi.org/10.1038/sj.ki.5000262.

Fukuyama, F. 2002. *Our Posthuman Future*. NY: Farrar, Straus and Giroux.

Fukuyama, F. 2004. Transhumanism. *Foreign Policy* (144): 42–43.

Fukuyama, Francis. 2002. Life, but Not as We Know It. *New Scientist (1971)* 174 (2339): 42–45.

Gallimore, Andrew R. 2015. Restructuring Consciousness—The Psychedelic State in Light of Integrated Information Theory. *Frontiers in Human Neuroscience* 9 (June): 346. https://doi.org/10.3389/fnhum.2015.00346.

Gampel, Eric. 2006. Does Professional Autonomy Protect Medical Futility Judgments? *Bioethics* 20 (2): 92–104.

Garcia-Santibanez, Rocio, and Harini Sarva. 2015. Isolated Hyperreligiosity in a Patient with Temporal Lobe Epilepsy. *Case Reports in Neurological Medicine* 2015 (August): 235856. https://doi.org/10.1155/2015/235856.

García-Velasco, J.A., and A. Pellicer. 2007. Cancer Treatment and Fertility: a Time to Reassess Realistic Opportunities. *Clinical & Translational Oncology* 9 (11): 681–682.

Gardiner, P. 2003. A Virtue Ethics Approach to Moral Dilemmas in Medicine. *Journal of Medical Ethics* 29 (5): 297–302. https://doi.org/10.1136/jme.29.5.297.

Garnett, Alex, Louise Whiteley, Heather Piwowar, Edie Rasmussen, and Judy Illes. 2011. Neuroethics and fMRI: Mapping a Fledgling Relationship. *Plos One* 6 (4): e18537. https://doi.org/10.1371/journal.pone.0018537.

Gensler, Harry J. 1985. A Kantian Argument Against Abortion. *Philosophical Studies* 48 (1): 57–72.

Geschwind, N. 1979. Behavioural Changes in Temporal Lobe Epilepsy. *Psychological Medicine* 9 (2): 217–219.

Gibson, Daniel G., John I. Glass, Carole Lartigue, Vladimir N. Noskov, Ray-Yuan Chuang, Mikkel A. Algire, Gwynedd A. Benders, et al. 2010. Creation of a Bacterial Cell Controlled by a Chemically Synthesized Genome. *Science* 329 (5987): 52–56. https://doi.org/10.1126/science.1190719.

Gibson, T.M. 2006. The Bioethics of Enhancing Human Performance for Spaceflight. *Journal of Medical Ethics* 32 (3): 129–132. https://doi.org/10.1136/jme.2005.012534.

Gilbert, David M. 2004. The Future of Human Embryonic Stem Cell Research: Addressing Ethical Conflict with Responsible Scientific Research. *Medical Science Monitor* 10 (5): RA99–103.

Giles, Jim. 2004. Nanotech Takes Small Step Towards Burying 'Grey Goo'. *Nature* 429 (6992): 591. https://doi.org/10.1038/429591b.

Gillon, R. 1985. Utilitarianism. *British Medical Journal (Clinical Research Ed.)* 290 (6479): 1411–1413.

Gillon, R. 1996. Brain Transplantation, Personal Identity and Medical Ethics. *Journal of Medical Ethics* 22 (3): 131–132.

Goldstein, L.B., and J.N. Davis. 1990. Influence of Lesion Size and Location on Amphetamine-facilitated Recovery of Beam-walking in Rats. *Behavioral Neuroscience* 104 (2): 320–327.

Gouzoulis-Mayfrank, E., and J. Daumann. 2006. Neurotoxicity of Methylenedioxyamphetamines (MDMA; Ecstasy) in Humans: How Strong Is the Evidence for Persistent Brain Damage? *Addiction* 101 (3): 348–361. https://doi.org/10.1111/j.1360-0443.2006.01314.x.

Graber, M.A. 2006. *Dred Scott and the Problem of Constitutional Evil*. Cambridge Studies on the American Constitution. New York: Cambridge University Press.

Granqvist, Pehr, Mats Fredrikson, Patrik Unge, Andrea Hagenfeldt, Sven Valind, Dan Larhammar, and Marcus Larsson. 2005. Sensed Presence and Mystical Experiences Are Predicted by Suggestibility, Not by the Application of Transcranial Weak Complex Magnetic Fields. *Neuroscience Letters* 379 (1): 1–6.

Granqvist, Pehr, Mario Mikulincer, and Phillip R Shaver. 2010. "Religion as Attachment: Normative Processes and Individual Differences." *Personality and Social Psychology Review : An Official Journal of the Society for Personality and Social Psychology, Inc* 14 (1): 49–59. https://doi.org/10.1177/1088868309348618.

Greek, Ray, Annalea Pippus, and Lawrence A. Hansen. 2012. The Nuremberg Code Subverts Human Health and Safety by Requiring Animal Modeling. *BMC Medical Ethics* 13 (July): 16. https://doi.org/10.1186/1472-6939-13-16.

Griffiths, R.R., W.A. Richards, U. McCann, and R. Jesse. 2006. Psilocybin Can Occasion Mystical-type Experiences Having Substantial and Sustained Personal Meaning and Spiritual Significance. *Psychopharmacology* 187 (3): 268–83, discussion 284. https://doi.org/10.1007/s00213-006-0457-5.

Gunderson, Martin. 2004. A Kantian View of Suicide and End-of-life Treatment. *Journal of Social Philosophy* 35 (2): 277–287.

Gutierrez, Roberto, and Roger Giner-Sorolla. 2007. Anger, Disgust, and Presumption of Harm as Reactions to Taboo-breaking Behaviors. *Emotion* 7 (4): 853–868. https://doi.org/10.1037/1528-3542.7.4.853.

Guttinger, Stephan. 2017. Trust in Science: CRISPR-Cas9 and the Ban on Human Germline Editing. *Science and Engineering Ethics*, June. https://doi.org/10.1007/s11948-017-9931-1.

Halevy, A., and B. Brody. 1993. Brain Death: Reconciling Definitions, Criteria, and Tests. *Annals of Internal Medicine* 119 (6): 519–525.

Hall, Wayne, and Michael Lynskey. 2009. The Challenges in Developing a Rational Cannabis Policy. *Current Opinion in Psychiatry* 22 (3): 258–262. https://doi.org/10.1097/YCO.0b013e3283298f36.

Haller, József, and Menno R. Kruk. 2006. Normal and Abnormal Aggression: Human Disorders and Novel Laboratory Models. *Neuroscience and Biobehavioral Reviews* 30 (3): 292–303. https://doi.org/10.1016/j.neubiorev.2005.01.005.

Harnett, M.J., A.D. Miller, R.J. Hurley, and K. Bhavani-Shankar. 2000. Pregnancy, Labour and Delivery in a Jehovah's Witness with Esophageal Varices and Thrombocytopenia. *Canadian Journal of Anaesthesia* 47 (12): 1253–1255. https://doi.org/10.1007/BF03019877.

Harold, F.M. 2001. *The Way of the Cell: Molecules, Organisms and the Order of Life*. Oxford, UK: Oxford University Press.

Harris, J. 1998. *Clones, Genes, and Immortality*. Oxford, UK: Oxford University Press.

Harris, J. 2003. In Praise of Unprincipled Ethics. *Journal of Medical Ethics* 29 (5): 303–306.

Harris, J. 2004. *On Cloning*. New York: Routledge.

Harris, J. 2007. *Enhancing Evolution*. Princeton NJ: Princeton University Press.

Harris, John. 2005a. Scientific Research Is a Moral Duty. *Journal of Medical Ethics* 31 (4): 242–248. https://doi.org/10.1136/jme.2005.011973.

Harris, John. 2005b. Sex Selection and Regulated Hatred. *Journal of Medical Ethics* 31 (5): 291–294. https://doi.org/10.1136/jme.2003.007526.

Harris, John. 2005c. No Sex Selection Please, We're British. *Journal of Medical Ethics* 31 (5): 286–288. https://doi.org/10.1136/jme.2004.008870.

Harris, John. 2009. Is It Acceptable for People to Take Methylphenidate to Enhance Performance? Yes. *BMJ (Clinical Research Ed.)* 338 (June): b1955. https://doi.org/10.1136/bmj.b1955.

Harris, John. 2011. Moral Enhancement and Freedom. *Bioethics* 25 (2): 102–111. https://doi.org/10.1111/j.1467-8519.2010.01854.x.

Harris, S. 1992. Japanese Biological Warfare Research on Humans: A Case Study of Microbiology and Ethics. *Annals of the New York Academy of Sciences* 666 (December): 21–52.

Haupt, W.F., and J. Rudolf. 1999. European Brain Death Codes: A Comparison of National Guidelines. *Journal of Neurology* 246 (6): 432–437.

Heinrichs, Markus, Bernadette von Dawans, and Gregor Domes. 2009. Oxytocin, Vasopressin, and Human Social Behavior. *Frontiers in Neuroendocrinology* 30 (4): 548–557. https://doi.org/10.1016/j.yfrne.2009.05.005.

Hershenov, David. 2003. The Problematic Role of 'Irreversibility' in the Definition of Death. *Bioethics* 17 (1): 89–100.

Heubel, Friedrich, and Nikola Biller-Andorno. 2005. The Contribution of Kantian Moral Theory to Contemporary Medical Ethics: A Critical Analysis. *Medicine, Health Care, and Philosophy* 8 (1): 5–18.

Howard, Joseph. 2005. The Moral Status of the Human Embryo According to Peter Singer: Individuality, Humanity, and Personhood. *The Linacre Quarterly* 72 (3): 212–228.

Huband, Nick, Michael Ferriter, Rajan Nathan, and Hannah Jones. 2010. Antiepileptics for Aggression and Associated Impulsivity. *Cochrane Database of Systematic Reviews* (2) (February): CD003499. https://doi.org/10.1002/14651858.cd003499.pub3.

Huff, Wolfgang, Doris Lenartz, Michael Schormann, Sun-Hee Lee, Jens Kuhn, Anastosious Koulousakis, Juergen Mai, et al. 2010. Unilateral Deep Brain Stimulation of the Nucleus Accumbens in Patients with Treatment-resistant Obsessive-compulsive Disorder: Outcomes after One Year. *Clinical Neurology and Neurosurgery* 112 (2): 137–143. https://doi.org/10.1016/j.clineuro.2009.11.006.

Hug, Kristina. 2006. Therapeutic Perspectives of Human Embryonic Stem Cell Research Versus the Moral Status of a Human Embryo—Does One Have to Be Compromised for the Other? *Medicina (Kaunas, Lithuania)* 42 (2): 107–114.

Hughes, James. 2004. *Citizen Cyborg: Why Democratic Societies Must Respond to the Redesigned Human of the Future*. Cambridge, MA: Westview Press.

Huh, Ryoong, In Bo Han, MoonYoung Chung, and Sangsup Chung. 2010. Comparison of Treatment Results Between Selective Peripheral Denervation and Deep Brain Stimulation in Patients with Cervical Dystonia. *Stereotactic and Functional Neurosurgery* 88 (4): 234–238. https://doi.org/10.1159/000314359.

Huxley, A. 1954. *The Doors of Perception*. London: Chatto & Windus.

Ikemoto, S., and J. Panksepp. 1999. The Role of Nucleus Accumbens Dopamine in Motivated Behavior: A Unifying Interpretation with Special Reference to Reward-seeking. *Brain Research Reviews* 31 (1): 6–41. https://doi.org/10.1016/S0165-0173(99)00023-5.

Iltis, Ana Smith, and Mark J. Cherry. 2010. Death Revisited: Rethinking Death and the Dead Donor Rule. *The Journal of Medicine and Philosophy* 35 (3): 223–241. https://doi.org/10.1093/jmp/jhq017.

Insel, T.R. 1997. A Neurobiological Basis of Social Attachment. *The American Journal of Psychiatry* 154 (6): 726–735. https://doi.org/10.1176/ajp.154.6.726.

Insel, Thomas R. 2010. The Challenge of Translation in Social Neuroscience: A Review of Oxytocin, Vasopressin, and Affiliative Behavior. *Neuron* 65 (6): 768–779. https://doi.org/10.1016/j.neuron.2010.03.005.

Institute of Medicine (US). 2000. Committee on Quality of Health Care in America. In*To Err is Human: Building a Safer Health System*, Eds. Kohn LT, Corrigan JM, Donaldson MS. Washington (DC): National Academies Press (US). (PubMed PMID: 25077248).

Inzlicht, Michael, Ian McGregor, Jacob B. Hirsh, and Kyle Nash. 2009. Neural Markers of Religious Conviction. *Psychological Science* 20 (3): 385–392. https://doi.org/10.1111/j.1467-9280.2009.02305.x.

James, Laura P., Philip R. Mayeux, and Jack A. Hinson. 2003. Acetaminophen-induced Hepatotoxicity. *Drug Metabolism and Disposition: The Biological Fate of Chemicals* 31 (12): 1499–1506. https://doi.org/10.1124/dmd.31.12.1499.

Jamieson, D. 1999. *Singer and His Critics*. Malden, MA: Blackwell.

Johnson, Jane. 2013. Vulnerable Subjects? The Case of Nonhuman Animals in Experimentation. *Journal of Bioethical Inquiry* 10 (4): 497–504. https://doi.org/10.1007/s11673-013-9473-4.

Jones, G.E. 1984. Fetal Brain Waves and Personhood. *Journal of Medical Ethics* 10 (4): 216–217.

Jonsen, Albert R. 1995. Casuistry: An Alternative or Complement to Principles? *Kennedy Institute of Ethics Journal* 5 (3): 237–251.

Kaempffert, W. 1943. Process of Dying. *New York Times*, April 10.

Kahane, Philippe, and Antoine Depaulis. 2010. Deep Brain Stimulation in Epilepsy: What Is Next? *Current Opinion in Neurology* 23 (2): 177–182. https://doi.org/10.1097/WCO.0b013e3283374a39.

Kamm, Frances M. 2005. Ethical Issues in Using and Not Using Embryonic Stem Cells. *Stem Cell Reviews* 1 (4): 325–330. https://doi.org/10.1385/SCR:1:4:325.

Kanai, Ryota, Tom Feilden, Colin Firth, and Geraint Rees. 2011. Political Orientations Are Correlated with Brain Structure in Young Adults. *Current Biology* 21 (8): 677–680. https://doi.org/10.1016/j.cub.2011.03.017.

Kao, Audiey C., and Kayhan P. Parsi. 2004. Content Analyses of Oaths Administered at U.S. Medical Schools in 2000. *Academic Medicine* 79 (9): 882–887.

Kapogiannis, Dimitrios, Aron K. Barbey, Su Michael, Giovanna Zamboni, Frank Krueger, and Jordan Grafman. 2009. Cognitive and Neural Foundations of Religious Belief. *Proceedings of the National Academy of Sciences of the United States of America* 106 (12): 4876–4881. https://doi.org/10.1073/pnas.0811717106.

Kass, L.R. 1997. The Wisdom of Repugnance. *The New Republic* (216): 17–26.

Kattera, Suresh, and Christopher Chen. 2006. Cryopreservation of Embryos by Vitrification: Current Development. *International Surgery* 91 (5 Suppl): S55–S62.

Katz, J. 1996. The Nuremberg Code and the Nuremberg Trial. A Reappraisal. *The Journal of the American Medical Association* 276 (20): 1662–1666.

Katz, Ralph V., B. Lee Green, Nancy R. Kressin, S. Stephen Kegeles, Min Qi Wang, Sherman A. James, Stefanie L. Russell, Cristina Claudio, and Jan M. McCallum. 2008. The Legacy of the Tuskegee Syphilis Study: Assessing Its Impact on Willingness to Participate in Biomedical Studies. *Journal of Health Care for the Poor and Underserved* 19 (4): 1168–1180. https://doi.org/10.1353/hpu.0.0067.

Katz, Ralph V., S. Stephen Kegeles, Nancy R. Kressin, B. Lee Green, Sherman A. James, Min Qi Wang, Stefanie L. Russell, and Cristina Claudio. 2008. Awareness of the Tuskegee Syphilis Study and the US Presidential Apology and Their Influence on Minority Participation in Biomedical Research. *American Journal of Public Health* 98 (6): 1137–1142. https://doi.org/10.2105/AJPH.2006.100131.

Keller, Charles F. 2003. Global Warming: The Balance of Evidence and Its Policy Implications. A Review of the Current State-of-the-controversy. *The Scientific World Journal* 3 (May): 357–411. https://doi.org/10.1100/tsw.2003.26.

Keller, Charles F. 2007. Global Warming 2007. An Update to Global Warming: The Balance of Evidence and Its Policy Implications. *The Scientific World Journal* 7 (March): 381–399. https://doi.org/10.1100/tsw.2007.91.

Kenagy, David N., Christopher T. Bird, Christopher M. Webber, and Joseph R. Fischer. 2004. Dextroamphetamine Use During B-2 Combat Missions. *Aviation, Space, and Environmental Medicine* 75 (5): 381–386.

Kevles, D.J. 1985. *In the Name of Eugenics. Genetics and the Uses of Human Heredity*. Berkeley, CA: University of California Press.

Kim, C-Yoon, Hanseul Oh, Xiaoping Ren, and Sergio Canavero. 2017. Immunohistochemical Evidence of Axonal Regrowth Across Polyethylene Glycol-fused Cervical Cords in Mice. *Neural Regeneration Research* 12 (1): 149–150. https://doi.org/10.4103/1673-5374.199014.

Kimberly, Michael. 2002. Reevaluating Repugnance: a Critical Analysis of Leon Kass' Writings on Genetic Reproductive Technologies. *Princeton Journal of Bioethics* 5: 8–24.

King, B.J. 2007. *Evolving God*. New York, NY: Doubleday.

Kious, B.M. 2001. The Nuremberg Code: Its History and Implications. *Princeton Journal of Bioethics* 4: 7–19.

Kirkpatrick, Lee A. 2004. *Attachment, Evolution, and the Psychology of Religion*. New York: Guilford Press.

Knutson, B., O.M. Wolkowitz, S.W. Cole, T. Chan, E.A. Moore, R.C. Johnson, J. Terpstra, R.A. Turner, and V.I. Reus. 1998. Selective Alteration of Personality and Social Behavior by Serotonergic Intervention. *The American Journal of Psychiatry* 155 (3): 373–379. https://doi.org/10.1176/ajp.155.3.373.

Koberda, J.L., W.M. Clark, H. Lutsep, and G. Nesbit. 1997. Successful Clinical Recovery and Reversal Ofmid-basilar Occlusion in Clinically Brain Dead Patient with Intra-arterialurokinase. *Neurology* 48 (suppl): A154.

Kohlberg, L. 1971. Stages in Moral Development as a Basis for Moral Education. In *Moral Education: Interdisciplinary Approaches*, ed. C.M. Beck, B.S. Crittenden, and E.V. Sullivan. Toronto: Toronto University Press.

Kolber, Adam. 2011. Neuroethics: Give Memory-altering Drugs a Chance. *Nature* 476 (7360): 275–276. https://doi.org/10.1038/476275a.

Kolibianakis, Efstratios M., Christos A. Venetis, and Basil C. Tarlatzis. 2009. Cryopreservation of Human Embryos by Vitrification or Slow Freezing: Which One Is Better? *Current Opinion in Obstetrics & Gynecology* 21 (3): 270–274. https://doi.org/10.1097/GCO.0b013e3283297dd6.

Kosko, B. 1993. *Fuzzy Thinking*. New York: Hyperion.

Kraemer, Felicitas. 2011. Authenticity Anyone? The Enhancement of Emotions via Neuro-Psychopharmacology. *Neuroethics* 4 (1): 51–64. https://doi.org/10.1007/s12152-010-9075-3.

Kramer, P.D. 1993. *Listening to Prozac*. NY: Viking Press.

Krasnova, Irina N., Zuzana Justinova, Bruce Ladenheim, Subramaniam Jayanthi, Michael T. McCoy, Chanel Barnes, John E. Warner, Steven R. Goldberg, and Jean Lud Cadet. 2010. Methamphetamine Self-administration Is Associated with Persistent Biochemical Alterations in Striatal and Cortical Dopaminergic Terminals in the Rat. *Plos One* 5 (1): e8790. https://doi.org/10.1371/journal.pone.0008790.

Kuchta, J., A. Koulousakis, and V. Sturm. 2007. Neurosurgical Pain Therapy with Epidural Spinal Cord Stimulation (SCS). *Acta Neurochirurgica. Supplement* 97 (Pt 1): 65–70.

Lai, C.S., S.E. Fisher, J.A. Hurst, F. Vargha-Khadem, and A.P. Monaco. 2001. A Forkhead-domain Gene Is Mutated in a Severe Speech and Language Disorder. *Nature* 413 (6855): 519–523. https://doi.org/10.1038/35097076.

Landsborough, D. 1987. St. Paul and Temporal Lobe Epilepsy. *Journal of Neurology, Neurosurgery, and Psychiatry* 50 (6): 659–664.

Lanni, Cristina, Silvia C. Lenzken, Alessia Pascale, Igor Del Vecchio, Marco Racchi, Francesca Pistoia, and Stefano Govoni. 2008. Cognition Enhancers Between Treating and Doping the Mind. *Pharmacological Research* 57 (3): 196–213. https://doi.org/10.1016/j.phrs.2008.02.004.

Lanphier, Edward, Fyodor Urnov, Sarah Ehlen Haecker, Michael Werner, and Joanna Smolenski. 2015. Don't Edit the Human Germ Line. *Nature* 519 (7544): 410–411. https://doi.org/10.1038/519410a.

Larson, Anne M. 2007. Acetaminophen Hepatotoxicity. *Clinics in Liver Disease* 11 (3): 525–48, vi. https://doi.org/10.1016/j.cld.2007.06.006.

Larson, Paul Sloan. 2008. Deep Brain Stimulation for Psychiatric Disorders. *Neurotherapeutics* 5 (1): 50–58. https://doi.org/10.1016/j.nurt.2007.11.006.

Launis, Veikko. 2010. Cosmetic Neurology: Sliding down the Slippery Slope? *Cambridge Quarterly of Healthcare Ethics: CQ: The International Journal of Healthcare Ethics Committees* 19 (2): 218–229. https://doi.org/10.1017/S096318010999048X.

Lawlor, Rob. 2007. Moral Theories in Teaching Applied Ethics. *Journal of Medical Ethics* 33 (6): 370–372. https://doi.org/10.1136/jme.2006.018044.

Leask, Julie. 2015. Should We Do Battle with Antivaccination Activists? *Public Health Research & Practice* 25 (2): e2521515. https://doi.org/10.17061/phrp2521515.

Lee, Tim W.R, and Kevin W. Southern. 2013. Topical Cystic Fibrosis Transmembrane Conductance Regulator Gene Replacement for Cystic Fibrosis-related Lung Disease. *Cochrane Database of Systematic Reviews* (11) (November): CD005599. https://doi.org/10.1002/14651858.cd005599.pub4.

Lemaire, François. 2006. The Nuremberg Doctors' Trial: The 60th Anniversary. *Intensive Care Medicine* 32 (12): 2049–2052. https://doi.org/10.1007/s00134-006-0467-0.

Letsoalo, J.L. 1998. Law, Blood Transfusions and Jehovah's Witnesses. *Medicine and Law* 17 (4): 633–638.

Levy, Neil. 2011. Neuroethics: A New Way of Doing Ethics. *AJOB Neuroscience* 2 (2): 3–9. https://doi.org/10.1080/21507740.2011.557683.

Levy, Robert, Timothy R. Deer, and Jaimie Henderson. 2010. Intracranial Neurostimulation for Pain Control: A Review. *Pain Physician* 13 (2): 157–165.

Li, Zhi, Wei Zhao, Wei Liu, Ye Zhou, Jingqiao Jia, and Lifeng Yang. 2014. Transplantation of Placenta-derived Mesenchymal Stem Cell-induced Neural Stem Cells to Treat Spinal Cord Injury. *Neural Regeneration Research* 9 (24): 2197–2204. https://doi.org/10.4103/1673-5374.147953.

Li, Peng-Wei, Xin Zhao, Yun-Long Zhao, Bing-Jian Wang, Yang Song, Zi-Long Shen, Hong-Jun Jiang, Hai Jin, Sergio Canavero, and Xiao-Ping Ren. 2017. A Cross-circulated Bicephalic Model of Head Transplantation. *CNS Neuroscience & Therapeutics* 23 (6): 535–541. https://doi.org/10.1111/cns.12700.

Lifshitz-Ben-Basat, Adi, and Nira Mashal. 2018. Improving Naming Abilities Among Healthy Young-Old Adults Using Transcranial Direct Current Stimulation. *Journal of Psycholinguistic Research* 47 (1): 113–124. https://doi.org/10.1007/s10936-017-9516-9.

Lifton, R.J. 1986. *The Nazi Doctors, Medical Killing & the Psychology of Genocide*. NY: Basic Books.

Lin, L., K.J. Vicente, and D.J. Doyle. 2001. Patient Safety, Potential Adverse Drug Events, and Medical Device Design: a Human Factors Engineering Approach. *Journal of Biomedical Informatics* 34 (4): 274–284. https://doi.org/10.1006/jbin.2001.1028.

Loutradi, Kalliopi E., Efstratios M. Kolibianakis, Christos A. Venetis, Evangelos G. Papanikolaou, George Pados, Ioannis Bontis, and Basil C. Tarlatzis. 2008. Cryopreservation of Human Embryos by Vitrification or Slow Freezing: A Systematic Review and Meta-analysis. *Fertility and Sterility* 90 (1): 186–193. https://doi.org/10.1016/j.fertnstert.2007.06.010.

Luce, J.M. 1995. Physicians Do Not Have a Responsibility to Provide Futile or Unreasonable Care If a Patient or Family Insists. *Critical Care Medicine* 23 (4): 760–766.

Lynch, Gary, and Christine M. Gall. 2006. Ampakines and the Threefold Path to Cognitive Enhancement. *Trends in Neurosciences* 29 (10): 554–562. https://doi.org/10.1016/j.tins.2006.07.007.

Machado, C. 1994. Death on Neurological Grounds. *Journal of Neurosurgical Sciences* 38 (4): 209–222.

Machado, Calixto, and Gerry Leisman. 2009. Towards an Effective Definition of Death and Disorders of Consciousness. *Reviews in the Neurosciences* 20 (3–4): 147–150.

Maleck, W.H., S.N. Piper, J. Triem, J. Boldt, and F.U. Zittel. 1998. Unexpected Return of Spontaneous Circulation after Cessation of Resuscitation (Lazarus Phenomenon). *Resuscitation* 39 (1–2): 125–128.

Maltz, E.M. 2007. *Dred Scott and the Politics of Slavey. Landmark Law Cases & American Society*. Lawrence, Kansas: University Press of Kansas.

Mammen, Biju, T. Ramakrishnan, Uma Sudhakar, and Vijayalakshmi. 2007. Principles of Gene Therapy. *Indian Journal of Dental Research : Official Publication of Indian Society for Dental Research* 18 (4): 196–200.

Mann, M.C., J. Votto, J. Kambe, and M.J. McNamee. 1992. Management of the Severely Anemic Patient Who Refuses Transfusion: Lessons Learned During the Care of a Jehovah's Witness. *Annals of Internal Medicine* 117 (12): 1042–1048.

Mantione, Mariska, Wim van de Brink, P. Richard Schuurman, and Damiaan Denys. 2010. Smoking Cessation and Weight Loss after Chronic Deep Brain Stimulation of the Nucleus Accumbens: Therapeutic and Research Implications: Case Report. *Neurosurgery* 66 (1): E218; discussion E218. https://doi.org/10.1227/01.neu.0000360570.40339.64.

Marazziti, D., H.S. Akiskal, A. Rossi, and G.B. Cassano. 1999. Alteration of the Platelet Serotonin Transporter in Romantic Love. *Psychological Medicine* 29 (3): 741–745.

Marshall, Ernest. 1986. Does the Moral Philosophy of the Belmont Report Rest on a Mistake? *IRB* 8 (6): 5–6.

Marx, K., and F. Engels. 2002. *On Religion*. Netherlands: Fredonia Books.

McIntyre, Alison. 2004. The Double Life of Double Effect. *Theoretical Medicine and Bioethics* 25 (1): 61–74.

McNally, J. Dayre, Christopher M. Sturgeon, and Kenneth B. Storey. 2003. Freeze-induced Expression of a Novel Gene, Fr47, in the Liver of the Freeze-tolerant Wood Frog, Rana Sylvatica. *Biochimica et Biophysica Acta* 1625 (2): 183–191.

Melega, William P., Matthew J. Jorgensen, Goran Laćan, Baldwin M. Way, Jamie Pham, Grenvill Morton, Arthur K. Cho, and Lynn A. Fairbanks. 2008. Long-term Methamphetamine Administration in the Vervet Monkey Models Aspects of a Human Exposure: Brain Neurotoxicity and Behavioral Profiles. *Neuropsychopharmacology* 33 (6): 1441–1452. https://doi.org/10.1038/sj.npp.1301502.

Merkle, R.C. 1992. The Technical Feasibility of Cryonics. *Medical Hypotheses* 39 (1): 6–16.

Metzger, T. 1998. *The Birth of Heroin & the Demonization of the Dope Fiend*. Port Townsend, WA: Loompanics Unlimited.

Meyerlindenberg, A. 2008. Impact of Prosocial Neuropeptides on Human Brain Function. In *Advances in Vasopressin and Oxytocin—From Genes to Behaviour to Disease*, 463–470. Elsevier. https://doi.org/10.1016/s0079-6123(08)00436-6.

Mickelsen, Ruth A., Daniel S. Bernstein, Mary Faith Marshall, and Steven H. Miles. 2013. The Barnes Case: Taking Difficult Futility Cases Public. *The Journal of Law, Medicine & Ethics : a Journal of the American Society of Law, Medicine & Ethics* 41 (1): 374–378. https://doi.org/10.1111/jlme.12027.

Mikhail, John. 2007. Universal Moral Grammar: Theory, Evidence and the Future. *Trends in Cognitive Sciences* 11 (4): 143–152. https://doi.org/10.1016/j.tics.2006.12.007.

Miller, R.B. 2002. How the Belmont Report Fails. *Essays in Philosophy* 4 (2): Article 6.

Mohandas, A., and S.N. Chou. 1971. Brain Death. A Clinical and Pathological Study. *Journal of Neurosurgery* 35 (2): 211–218. https://doi.org/10.3171/jns.1971.35.2.0211.

Mohandas, E. 2008. Neurobiology of Spirituality. *Mens Sana Monographs* 6 (1): 63–80. https://doi.org/10.4103/0973-1229.33001.

Mokwa, Wilfried. 2007. An Implantable Microsystem as a Vision Prosthesis. *Medical Device Technology* 18 (6): 20.

Moll, Jorge, Ricardo de Oliveira-Souza, Fernanda Tovar Moll, Fátima Azevedo Ignácio, Ivanei E. Bramati, Egas M. Caparelli-Dáquer, and Paul J. Eslinger. 2005. The Moral Affiliations of Disgust: a Functional MRI Study. *Cognitive and Behavioral Neurology: Official Journal of the Society for Behavioral and Cognitive Neurology* 18 (1): 68–78.

Mondello, Sarah E., Michael R. Kasten, Philip J. Horner, and Chet T. Moritz. 2014. Therapeutic Intraspinal Stimulation to Generate Activity and Promote Long-term Recovery. *Frontiers in Neuroscience* 8 (February): 21. https://doi.org/10.3389/fnins.2014.00021.

Morewedge, C. K. and M.I Norton. 2009. When dreaming is believing: The (motivated) interpretation of dreams .*JPers Soc Psychol*. 96(2): 249–64.

Morison, RS. 1971. Death: Process or event? *Science*. 173 (3998): 694–698. (PubMed PMID: 5568501).

Morelli, M. 1999. Commerce in Organs: a Kantian Critique. *Journal of Social Philosophy* 30 (2): 315–324.

Morris, Kelly. 2008. Experts Urge Smart Thinking on Cognitive Enhancers. *Lancet Neurology* 7 (6): 476–477. https://doi.org/10.1016/S1474-4422(08)70101-7.

Mosher, J.F., and K.L. Yanagisako. 1991. Public Health, Not Social Warfare: A Public Health Approach to Illegal Drug Policy. *Journal of Public Health Policy* 12 (3): 278–323.

Muir, Timothy J., Jon P. Costanzo, and Richard E. Lee. 2007. Osmotic and Metabolic Responses to Dehydration and Urea-loading in a Dormant, Terrestrially Hibernating Frog. *Journal of Comparative Physiology. B, Biochemical, Systemic, and Environmental Physiology* 177 (8): 917–926. https://doi.org/10.1007/s00360-007-0190-3.

Muramoto, O. 1998. Bioethics of the Refusal of Blood by Jehovah's Witnesses: Part 1. Should Bioethical Deliberation Consider Dissidents' Views? *Journal of Medical Ethics* 24 (4): 223–230.

Muramoto, O. 1998. Bioethics of the Refusal of Blood by Jehovah's Witnesses: Part 2. A Novel Approach Based on Rational Non-interventional Paternalism. *Journal of Medical Ethics* 24 (5): 295–301.

Muramoto, O. 1999. Bioethics of the Refusal of Blood by Jehovah's Witnesses: Part 3. A Proposal for a Don't-ask-don't-tell Policy. *Journal of Medical Ethics* 25 (6): 463–468.

Muramoto, Osamu. 2004. The Role of the Medial Prefrontal Cortex in Human Religious Activity. *Medical Hypotheses* 62 (4): 479–485. https://doi.org/10.1016/j.mehy.2003.10.010.

Murray, Robin M., Alessandra Paparelli, Paul D. Morrison, Arianna Marconi, and Marta Di Forti. 2013. What Can We Learn About Schizophrenia from Studying the Human Model, Drug-induced Psychosis? *American Journal of Medical Genetics. Part B, Neuropsychiatric Genetics* 162B (7): 661–670. https://doi.org/10.1002/ajmg.b.32177.

Nencini, Paolo. 2002. The Shaman and the Rave Party: Social Pharmacology of Ecstasy. *Substance Use & Misuse* 37 (8–10): 923–939.

Nencini, Paolo, and Kathleen A. Grant. 2010. Psychobiology of Drug-induced Religious Experience: From the Brain 'Locus of Religion' to Cognitive Unbinding. *Substance Use & Misuse* 45 (13): 2130–2151. https://doi.org/10.3109/10826081003713803.

Newberg, A., A. Alavi, M. Baime, M. Pourdehnad, J. Santanna, and E. d'Aquili. 2001. The Measurement of Regional Cerebral Blood Flow During the Complex Cognitive Task of Meditation: A Preliminary SPECT Study. *Psychiatry Research* 106 (2): 113–122.

Newberg, A.B., E.G. d'Aquili, and V. Rause. 2002. *Why God Won't Go Away: Brain Science and the Biology of Belief.* New York, NY: Ballantine Books.

Newberg, A.B., and M.R. Waldman. 2006. *Why We Believe What We Believe: Our Biological Need for Meaning, Spirituality, and Truth*. New York, NY: Free Press.

Ng, Felicity. 2007. The Interface Between Religion and Psychosis. *Australasian Psychiatry: Bulletin of Royal Australian and New Zealand College of Psychiatrists* 15 (1): 62–66. https://doi.org/10.1080/10398560601083118.

Nichols, David E. 2016. Psychedelics. *Pharmacological Reviews* 68 (2): 264–355. https://doi.org/10.1124/pr.115.011478.

Nie, Jing-Bao. 2006. The United States Cover-up of Japanese Wartime Medical Atrocities: Complicity Committed in the National Interest and Two Proposals for Contemporary Action. *The American Journal of Bioethics* 6 (3): W21–W33. https://doi.org/10.1080/15265160600686356.

Normann, Claus, and Mathias Berger. 2008. Neuroenhancement: Status Quo and Perspectives. *European Archives of Psychiatry and Clinical Neuroscience* 258 Suppl 5 (November): 110–114. https://doi.org/10.1007/s00406-008-5022-2.

Nuzzo, Jennifer B. 2006. The Biological Threat to U.S. Water Supplies: Toward a National Water Security Policy. *Biosecurity and Bioterrorism: Biodefense Strategy, Practice, and Science* 4 (2): 147–159. https://doi.org/10.1089/bsp.2006.4.147.

O'Connell, Garret. 2011. Tracking the Impact of Neuroethics. *Cortex* 47 (10): 1259–1260. https://doi.org/10.1016/j.cortex.2011.04.020.

Oliver, Joan D., and Randall S. Perry. 2006. Definitely Life but Not Definitively. *Origins of Life and Evolution of the Biosphere* 36 (5–6): 515–521. https://doi.org/10.1007/s11084-006-9035-4.

Olweus, D., A. Mattsson, D. Schalling, and H. Löw. 1988. Circulating Testosterone Levels and Aggression in Adolescent Males: A Causal Analysis. *Psychosomatic Medicine* 50 (3): 261–272.

Pagel, James F. 2012. What Physicians Need to Know About Dreams and Dreaming. *Current Opinion in Pulmonary Medicine* 18 (6): 574–579. https://doi.org/10.1097/MCP.0b013e328358bf42.

Pallis, C. 1980. Television and Brain Death. *British Medical Journal* 281 (6247): 1064.

Palmer, David C. 2006. On Chomsky's Appraisal of Skinner's Verbal Behavior: A Half Century of Misunderstanding. *The Behavior Analyst / MABA* 29 (2): 253–267.

Papsin, Blake C., and Karen A. Gordon. 2007. Cochlear Implants for Children with Severe-to-profound Hearing Loss. *The New England Journal of Medicine* 357 (23): 2380–2387. https://doi.org/10.1056/NEJMct0706268.

Parasuraman, R., and R.A. McKinley. 2014. Using Noninvasive Brain Stimulation to Accelerate Learning and Enhance Human Performance. *Human Factors: The Journal of the Human Factors and Ergonomics Society* 56 (5): 816–824. https://doi.org/10.1177/0018720814538815.

Parnas, Josef, and Mads Gram Henriksen. 2016. Mysticism and Schizophrenia: A Phenomenological Exploration of the Structure of Consciousness in the Schizophrenia Spectrum Disorders. *Consciousness and Cognition* 43 (July): 75–88. https://doi.org/10.1016/j.concog.2016.05.010.

Patterson, K.W., and A. McShane. 1991. Reflex Spinal Cord Activity as a Cause of a Delay in the Diagnosis of Brain Death. *Irish Medical Journal* 84 (1): 27–28.

Pegg, David E. 2007. Principles of Cryopreservation. *Methods in Molecular Biology* 368: 39–57. https://doi.org/10.1007/978-1-59745-362-2_3.

Pellegrino, E.D. 1995. Toward a Virtue-based Normative Ethics for the Health Professions. *Kennedy Institute of Ethics Journal* 5 (3): 253–277.

Penrose, R. 1989. *The Emperor's New Mind: Concerning Computers, Minds, and the Laws of Physics*. Oxford, UK: Oxford University Press.

Penrose, R. 1996. *Shadows of the Mind: A Search for the Missing Science of Consciousness*. Oxford, UK: Oxford University Press.

Peoples, Hervey C., and Frank W. Marlowe. 2012. Subsistence and the Evolution of Religion. *Human Nature (Hawthorne, N.Y.)* 23 (3): 253–269. https://doi.org/10.1007/s12110-012-9148-6.

Persinger, M.A. 1983. Religious and Mystical Experiences as Artifacts of Temporal Lobe Function: a General Hypothesis. *Perceptual and Motor Skills* 57 (3 Pt 2): 1255–1262. https://doi.org/10.2466/pms.1983.57.3f.1255.

Persinger, Michael A., and Faye Healey. 2002. Experimental Facilitation of the Sensed Presence: Possible Intercalation Between the Hemispheres Induced by Complex Magnetic Fields. *The Journal of Nervous and Mental Disease* 190 (8): 533–541. https://doi.org/10.1097/01.NMD.0000026653.84018.B0.

Pham, J.V., and T. Puzantian. 2001. Ecstasy: Dangers and Controversies. *Pharmacotherapy* 21 (12): 1561–1565.

Pommer, R.W. 1993. Donaldson V. Van de Kamp: Cryonics, Assisted Suicide, and the Challenges of Medical Science. *The Journal of Contemporary Health Law and Policy* 9: 589–603.

Porta, M., A. Brambilla, A.E. Cavanna, D. Servello, M. Sassi, H. Rickards, and M.M. Robertson. 2009. Thalamic Deep Brain Stimulation for Treatment-refractory Tourette Syndrome: Two-year Outcome. *Neurology* 73 (17): 1375–1380. https://doi.org/10.1212/WNL.0b013e3181bd809b.

President's Commission for the Study of Ethical Problemsin Medicine and Biomedical and Behavior Research. 1981. *Defining Death: A Report on the Medical, Legal and Ethical Issues in the Determination of Death*. Washington, DC: US Government Printing Office.

Puri, K.S., K.R. Suresh, N.J. Gogtay, and U.M. Thatte. 2009. Declaration of Helsinki, 2008: Implications for Stakeholders in Research. *Journal of Postgraduate Medicine* 55 (2): 131–134. https://doi.org/10.4103/0022-3859.52846.

Rawls, J. 1971. *A Theory of Justice*. Cambridge, MA: Belknap Press.

Ren, Shuai, Ze-Han Liu, Wu Qiong, Fu Kuang, Wu Jun, Li-Ting Hou, Ming Li, et al. 2017. Polyethylene Glycol-induced Motor Recovery after Total Spinal Transection in Rats. *CNS Neuroscience & Therapeutics* 23 (8): 680–685. https://doi.org/10.1111/cns.12713.

Ren, Xiaoping, and Michael C. Laugel. 2013. The Next Frontier in Composite Tissue Allotransplantation. *CNS Neuroscience & Therapeutics* 19 (1): 1–4. https://doi.org/10.1111/cns.12029.

Ren, Xiao-Ping, Yang Song, Yi-Jie Ye, Peng-Wei Li, Ke-Cheng Han, Zi-Long Shen, Ji-Gang Shan, Kristin Luther, and Bao-Feng Yang. 2014. Allogeneic Head and Body Reconstruction: Mouse Model. *CNS Neuroscience & Therapeutics* 20 (12): 1056–1060. https://doi.org/10.1111/cns.12341.

Richards, P.M., M.A. Persinger, and S.A. Koren. 1993. Modification of Activation and Evaluation Properties of Narratives by Weak Complex Magnetic Field Patterns That Simulate Limbic Burst Firing. *The International Journal of Neuroscience* 71 (1–4): 71–85.

Ridley, D.T. 1999. Jehovah's Witnesses' Refusal of Blood: Obedience to Scripture and Religious Conscience. *Journal of Medical Ethics* 25 (6): 469–472.

Romain, Tiffany. 2010. Extreme Life Extension: Investing in Cryonics for the Long, Long Term. *Medical Anthropology* 29 (2): 194–215. https://doi.org/10.1080/01459741003715391.

Rose, Gregory M., Voon S. Ong, and Diana S. Woodruff-Pak. 2007. Efficacy of MEM 1003, a Novel Calcium Channel Blocker, in Delay and Trace Eyeblink Conditioning in Older Rabbits. *Neurobiology of Aging* 28 (5): 766–773. https://doi.org/10.1016/j.neurobiolaging.2006.03.006.

Ross, W.D. 1930. *The Right and the Good*. Oxford, UK: Clarendon Press.

Rothenberg, David M. 1990. The Approach to the Jehovah's Witness Patient. *Anesthesiology Clinics of North America* 8 (3): 589–607.

Rowiński, Wojciech. 2007. Future of Transplantation Medicine. *Annals of Transplantation* 12 (1): 5–10.

Ryan, C.J. 1998. Pulling up the Runaway: The Effect of New Evidence on Euthanasia's Slippery Slope. *Journal of Medical Ethics* 24 (5): 341–344.

Sagoff, Mark. 2007. A Transcendental Argument for the Concept of Personhood in Neuroscience. *The American Journal of Bioethics* 7 (1): 72–3; discussion W1. https://doi.org/10.1080/15265160601064355.

Saini, P. 1999. The Doctrine of Double Effect and the Law of Murder. *The Medico-legal Journal* 67 (Pt 3): 106–120. https://doi.org/10.1258/rsmmlj.67.3.106.

Sandel, M.J. 2007. *The Case Against Perfection: Ethics in the Age of Genetic Engineering*. Cambridge, MA: Belknap Press.

Sanger, M.A. 1932. Plan for Peace" *The Birth Control Review.*

Saposnik, G., J.A. Bueri, J. Mauriño, R. Saizar, and N.S. Garretto. 2000. Spontaneous and Reflex Movements in Brain Death. *Neurology* 54 (1): 221–223.

Saposnik, Gustavo, Vincenzo S. Basile, and G. Bryan Young. 2009. Movements in Brain Death: A Systematic Review. *The Canadian Journal of Neurological Sciences. Le Journal Canadien Des Sciences Neurologiques* 36 (2): 154–160. https://doi.org/10.1017/s031716710000651x.

Saver, J.L., and J. Rabin. 1997. The Neural Substrates of Religious Experience. *The Journal of Neuropsychiatry and Clinical Neurosciences* 9 (3): 498–510. https://doi.org/10.1176/jnp.9.3.498.

Savulescu, J. 1995. Rational Non-interventional Paternalism: Why Doctors Ought to Make Judgments of What Is Best for Their Patients. *Journal of Medical Ethics* 21 (6): 327–331.

Savulescu, J. 1999. Should We Clone Human Beings? Cloning as a Source of Tissue for Transplantation. *Journal of Medical Ethics* 25 (2): 87–95.

Savulescu, J. 1999. Sex Selection: The Case For. *The Medical Journal of Australia* 171 (7): 373–375.

Savulescu, J. 2001. Why Genetic Testing for Genes for Criminality Is Morally Required. *Princeton Journal of Bioethics* 4: 79–97.

Savulescu, J. 2001. Procreative Beneficence: Why We Should Select the Best Children. *Bioethics* 15 (5–6): 413–426. https://doi.org/10.1111/1467-8519.00251.

Savulescu, J., and E. Dahl. 2000. Sex Selection and Preimplantation Diagnosis: A Response to the Ethics Committee of the American Society of Reproductive Medicine. *Human Reproduction* 15 (9): 1879–1880.

Savulescu, Julian. 2002. The Embryonic Stem Cell Lottery and the Cannibalization of Human Beings. *Bioethics* 16 (6): 508–529.

Savulescu, Julian, and Nick Bostrom. 2008. *Human Enhancement*. Oxford: Oxford University Press.

Sawyer, Robert J. 2007. Robot Ethics. *Science* 318 (5853): 1037. https://doi.org/10.1126/science.1151606.

Schaich Borg, Jana, Debra Lieberman, and Kent A Kiehl. 2008. Infection, Incest, and Iniquity: Investigating the Neural Correlates of Disgust and Morality. *Journal of Cognitive Neuroscience* 20 (9): 1529–1546. https://doi.org/10.1162/jocn.2008.20109.

Schermer, M.H.N. 2007. Brave New World Versus Island—Utopian and Dystopian Views on Psychopharmacology. *Medicine, Health Care, and Philosophy* 10 (2): 119–128. https://doi.org/10.1007/s11019-007-9059-1.

Schieman, S. 2010. Socioeconomic Status and Beliefs About God's Influence in Everyday Life. *Sociology of Religion* 71 (1): 25–51. https://doi.org/10.1093/socrel/srq004.

Schmidt, C.F. 1965. Pharmacological Support for Man on Long Space Flights. *Proceedings of the National Academy of Sciences of the United States of America* 53 (6): 1365–1369.

Schnall, Simone, Jonathan Haidt, Gerald L. Clore, and Alexander H. Jordan. 2008. Disgust as Embodied Moral Judgment. *Personality and Social Psychology Bulletin* 34 (8): 1096–1109. https://doi.org/10.1177/0146167208317771.

Schroeder, Joshua, Janina Kueper, Kaplan Leon, and Meir Liebergall. 2015. Stem Cells for Spine Surgery. *World Journal of Stem Cells* 7 (1): 186–194. https://doi.org/10.4252/wjsc.v7.i1.186.

Schwerkoske, John P., Jason P. Caplan, and Dawn M. Benford. 2012. Self-mutilation and Biblical Delusions: A Review. *Psychosomatics* 53 (4): 327–333. https://doi.org/10.1016/j.psym.2012.01.007.

Scott, David J., Christian S. Stohler, Christine M. Egnatuk, Heng Wang, Robert A. Koeppe, and Jon-Kar Zubieta. 2007. Individual Differences in Reward Responding Explain Placebo-induced Expectations and Effects. *Neuron* 55 (2): 325–336. https://doi.org/10.1016/j.neuron.2007.06.028.

Shander, A. 1999. Informed Consent and 'Reform' Jehovah's Witness Patients. *Anesthesiology* 90 (6): 1787.

Shanks, Niall, Ray Greek, and Jean Greek. 2009. Are Animal Models Predictive for Humans? *Philosophy, Ethics, and Humanities in Medicine: PEHM* 4 (January): 2. https://doi.org/10.1186/1747-5341-4-2.

Shapiro, M. 2001. Plasticity, Hippocampal Place Cells, and Cognitive Maps. *Archives of Neurology* 58 (6): 874–881.

Shapshay, Sandra, and Kenneth D. Pimple. 2007. Participation in Biomedical Research Is an Imperfect Moral Duty: A Response to John Harris. *Journal of Medical Ethics* 33 (7): 414–417. https://doi.org/10.1136/jme.2006.017384.

Shaw, David. 2009. Cryoethics: Seeking Life after Death. *Bioethics* 23 (9): 515–521. https://doi.org/10.1111/j.1467-8519.2009.01760.x.

Shemie, S.D. 2006. Severe Brain Injury to Neurological Determination of Death: Canadian Forum Recommendations. *Canadian Medical Association Journal* 174 (6): S1–S12. https://doi.org/10.1503/cmaj.045142.

Shermer, M. 2001. Nano Nonsense and Cryonics. *Scientific American* 285 (3): 29.

Sherwin, S. 1992. *No Longer Patient: Feminist Ethics and Health Care*. Philadelphia, PA: Temple University Press.

Shuster, E. 1997. Fifty Years Later: The Significance of the Nuremberg Code. *The New England Journal of Medicine* 337 (20): 1436–1440. https://doi.org/10.1056/NEJM199711133372006.

Singer, P. 1995. How Are We to Live? Ethics in an Age of Self-interest. Mandarin: London.

Singer, P.A. 1975. *Animal Liberation*. NY: Harper Collins.

Singer, P.A. 1996. *Rethinking Life and Death: The Collapse of Our Traditional Values*. New York: St. Martin's Griffin.

Singer, P.A. 2009. *The Life You Can Save: Acting Now to End World Poverty*. NY: Random House.

Singer, P.A. 2011. *Practical Ethics*, 2nd ed. Cambridge: Cambridge University Press.

Sinsheimer, Robert. 2004. Beyond Therapy …. *PLoS Biology* 2 (6): e181. https://doi.org/10.1371/journal.pbio.0020181.

Smith, G.P. 1983. The Iceperson Cometh: Cryonics, Law and Medicine. *Health Matrix* 1 (2): 23–35.

Smith, W.J. 2001. Revisiting the Belmont Report. *The Hastings Center Report* 31 (2): 5.

Sneiderman, Barney. 1991. The Shulman Case and the Right to Refuse Treatment. *Humane Medicine* 7 (1): 15–21.

Soeling, Caspar, and Eckart Voland. 2002. Toward an Evolutionary Psychology of Religiosity. *Neuro Endocrinology Letters* 23 Suppl 4 (December): 98–104.

Spiegel, Allen M. 2013. The Stem Cell Wars: A Dispatch from the Front. *Transactions of the American Clinical and Climatological Association* 124: 94–110.

Sreenivasan, Gopal. 2003. Does Informed Consent to Research Require Comprehension? *The Lancet* 362 (9400): 2016–2018. https://doi.org/10.1016/S0140-6736(03)15025-8.

Stagg, Charlotte J., and Michael A. Nitsche. 2011. Physiological Basis of Transcranial Direct Current Stimulation. *The Neuroscientist* 17 (1): 37–53. https://doi.org/10.1177/1073858410386614.

Starr, P.A., J.L. Vitek, and R.A. Bakay. 1998. Deep Brain Stimulation for Movement Disorders. *Neurosurgery Clinics of North America* 9 (2): 381–402.

Steeds, Hannah, Robin L. Carhart-Harris, and James M. Stone. 2015. Drug Models of Schizophrenia. *Therapeutic Advances in Psychopharmacology* 5 (1): 43–58. https://doi.org/10.1177/2045125314557797.

Stelten, Bianca M.L., Lieke H.M. Noblesse, Linda Ackermans, Yasin Temel, and Veerle Visser-Vandewalle. 2008. The Neurosurgical Treatment of Addiction. *Neurosurgical Focus* 25 (1): E5. https://doi.org/10.3171/FOC/2008/25/7/E5.

Stolberg, Sheryl Gay. 1999. The Biotech Death of Jesse Gelsinger. *The New York Times Magazine*, November, 14.

Storey, K.B., D.D. Mosser, D.N. Douglas, J.E. Grundy, and J.M. Storey. 1996. Biochemistry Below 0 Degrees C: Nature's Frozen Vertebrates. *Brazilian Journal of Medical and Biological Research* 29 (3): 283–307.

Storey, Kenneth B. 2004. Strategies for Exploration of Freeze Responsive Gene Expression: Advances in Vertebrate Freeze Tolerance. *Cryobiology* 48 (2): 134–145. https://doi.org/10.1016/j.cryobiol.2003.10.008.

St-Pierre, L.S., and M.A. Persinger. 2006. Experimental Facilitation of the Sensed Presence Is Predicted by the Specific Patterns of the Applied Magnetic Fields, Not by Suggestibility: Re-analyses of 19 Experiments. *The International Journal of Neuroscience* 116 (9): 1079–1096. https://doi.org/10.1080/00207450600808800.

Suckiel, Ellen. 2008. Human Embryonic Stem Cell Research: A Critical Survey of the Ethical Issues. *Advances in Pediatrics* 55: 79–96.

Swanson, George A. 2005. The Religious Objections and Military Opposition to Anesthetics, 1846–1848. *Bulletin of Anesthesia History* 23 (2): 1, 4–5, 14.

Tabakow, Pawel, Geoffrey Raisman, Wojciech Fortuna, Marcin Czyz, Juliusz Huber, Daqing Li, Pawel Szewczyk, et al. 2014. Functional Regeneration of Supraspinal Connections in a Patient with Transected Spinal Cord Following Transplantation of Bulbar Olfactory Ensheathing Cells with Peripheral Nerve Bridging. *Cell Transplantation* 23 (12): 1631–1655. https://doi.org/10.3727/096368914X685131.

Takala, Tuija. 2007. Concepts of 'Person' and 'Liberty', and Their Implications to Our Fading Notions of Autonomy. *Journal of Medical Ethics* 33 (4): 225–228. https://doi.org/10.1136/jme.2006.016816.

Tandy, C. 2002. *The Philosophy of Robert Ettinger*. Palo Alto, CA: Ria University Press.

Taylor, C. 1985. *Philosophical Papers: Volume 1, Human Agency and Language*, vol. 1. Cambridge, UK: Cambridge University Press.

Temme, Leonard A. 2003. Ethics in Human Experimentation: The Two Military Physicians Who Helped Develop the Nuremberg Code. *Aviation, Space, and Environmental Medicine* 74 (12): 1297–1300.

Thair, Hayley, Amy L. Holloway, Roger Newport, and Alastair D. Smith. 2017. Transcranial Direct Current Stimulation (tdcs): A Beginner's Guide for Design and Implementation. *Frontiers in Neuroscience* 11 (November): 641. https://doi.org/10.3389/fnins.2017.00641.

The American Academy of Neurology, Quality Standards Subcommittee. 1995. Quality Standards Subcommittee of the American Academy of Neurology. Practice Parameters for Determining Brain Death in Adults [summary Statement]. *Neurology* 45 (5): 1012–1014.

The President's Council on Bioethics. 2003. *Beyond Therapy: Biotechnology and the Pursuit of Human Improvement*. Washington, DC: US Government Printing Office.

Thomson, J.J. 1985. The Trolley Problem. *The Yale Law Journal* 94 (6): 1395–1415.

Treasure, T. 1998. Lessons from the Bristol Case. More Openness—On Risks and on Individual Surgeons' Performance. *BMJ (Clinical Research Ed.)* 316 (7146): 1685–1686.

Truog, R.D., and J.C. Fackler. 1992. Rethinking Brain Death. *Critical Care Medicine* 20 (12): 1705–1713.

Truog, Robert D., and Walter M. Robinson. 2003. Role of Brain Death and the Dead-donor Rule in the Ethics of Organ Transplantation. *Critical Care Medicine* 31 (9): 2391–2396. https://doi.org/10.1097/01.CCM.0000090869.19410.3C.

Tukmachev, Dmitry, Oleg Lunov, Vitalii Zablotskii, Alexandr Dejneka, Michal Babic, Eva Syková, and Šárka Kubinová. 2015. An Effective Strategy of Magnetic Stem Cell Delivery for Spinal Cord Injury Therapy. *Nanoscale* 7 (9): 3954–3958. https://doi.org/10.1039/c4nr05791k.

Tumpey, Terrence M., Christopher F. Basler, Patricia V. Aguilar, Hui Zeng, Alicia Solórzano, David E. Swayne, Nancy J. Cox, et al. 2005. Characterization of the Reconstructed 1918 Spanish Influenza Pandemic Virus. *Science* 310 (5745): 77–80. https://doi.org/10.1126/science.1119392.

Turing, A.M. 1950. Computing Machinery and Intelligence. *Mind* 59 (236): 433–460.

Turner, Leigh. 2003. Bioethics and Religions: Religious Traditions and Understandings of Morality, Health, and Illness. *Health Care Analysis: HCA: Journal of Health Philosophy and Policy* 11 (3): 181–197. https://doi.org/10.1023/B:HCAN.0000005491.88004.27.

Uppal, Talat, A. Pickering, K. Erasmus, J. Pardey, and Roy G. Beran. 2012. The Legal Status of the Fetus in New South Wales. *Journal of Law and Medicine* 20 (1): 178–183.

Vajta, G., Z.P. Nagy, A. Cobo, J. Conceicao, and J. Yovich. 2009. Vitrification in Assisted Reproduction: Myths, Mistakes, Disbeliefs and Confusion. *Reproductive Biomedicine Online* 19 (Suppl 3): 1–7.

Van den Belt, Henk. 2009. Playing God in Frankenstein's Footsteps: Synthetic Biology and the Meaning of Life. *Nanoethics* 3 (3): 257–268. https://doi.org/10.1007/s11569-009-0079-6.

Van Kuyck, K., L. Gabriëls, P. Cosyns, L. Arckens, V. Sturm, S. Rasmussen, and B. Nuttin. 2007. Behavioural and Physiological Effects of Electrical Stimulation in the Nucleus Accumbens: A Review. *Acta Neurochirurgica. Supplement* 97 (Pt 2): 375–391.

Van Norman, G.A. 1999. A matter of life and death: What every anesthesiologist should know about the medical, legal,and ethical aspects of declaring brain death. *Anesthesiology*. 91 (1): 275–87. (PubMed PMID: 10422953).

Veenema, Alexa H., and Inga D. Neumann. 2008. Central Vasopressin and Oxytocin Release: Regulation of Complex Social Behaviours. *Progress in Brain Research* 170: 261–276. https://doi.org/10.1016/S0079-6123(08)00422-6.

Vernes, Sonja C., Jérôme Nicod, Fanny M. Elahi, Julie A. Coventry, Niamh Kenny, Anne-Marie Coupe, Louise E. Bird, Kay E. Davies, and Simon E. Fisher. 2006. Functional Genetic Analysis of Mutations Implicated in a Human Speech and Language Disorder. *Human Molecular Genetics* 15 (21): 3154–3167. https://doi.org/10.1093/hmg/ddl392.

Veronelli, Laura, Sara J. Makaretz, Megan Quimby, Bradford C. Dickerson, and Jessica A. Collins. 2017. Geschwind Syndrome in Frontotemporal Lobar Degeneration: Neuroanatomical and Neuropsychological Features over 9 Years. *Cortex* 94 (June): 27–38. https://doi.org/10.1016/j.cortex.2017.06.003.

Viamonte, Sarah M., Karlene K. Ball, and Meredith Kilgore. 2006. A Cost-benefit Analysis of Risk-reduction Strategies Targeted at Older Drivers. *Traffic Injury Prevention* 7 (4): 352–359. https://doi.org/10.1080/15389580600791362.

Vicente, Kim J., Karima Kada-Bekhaled, Gillian Hillel, Andrea Cassano, and Beverley A. Orser. 2003. Programming Errors Contribute to Death from Patient-controlled Analgesia: Case Report and Estimate of Probability. *Canadian Journal of Anaesthesia* 50 (4): 328–332. https://doi.org/10.1007/BF03021027.

Walker-Batson, D., S. Curtis, R. Natarajan, J. Ford, N. Dronkers, E. Salmeron, J. Lai, and D.H. Unwin. 2001. A Double-blind, Placebo-controlled Study of the Use of Amphetamine in the Treatment of Aphasia. *Stroke* 32 (9): 2093–2098.

Ward, Simon E., Benjamin D. Bax, and Mark Harries. 2010. Challenges for and Current Status of Research into Positive Modulators of AMPA Receptors. *British Journal of Pharmacology* 160 (2): 181–190. https://doi.org/10.1111/j.1476-5381.2010.00726.x.

Watts, Jonathan. 2002. Victims of Japan's Notorious Unit 731 Sue. *The Lancet* 360 (9333): 628.

Weber, M. 1933. *The Sociology of Religion*. Boston: Beacon Press.

Weijer, C. 2000. Bioethics: An Anthology. *BMJ (Clinical Research Ed.)* 320 (7243): 1215A.

Weijer, C., P.A. Singer, B.M. Dickens, and S. Workman. 1998. Bioethics for Clinicians: 16. Dealing with Demands for Inappropriate Treatment. *Canadian Medical Association Journal* 159 (7): 817–821.

Wheatley, Thalia, and Jonathan Haidt. 2005. Hypnotic Disgust Makes Moral Judgments More Severe. *Psychological Science* 16 (10): 780–784. https://doi.org/10.1111/j.1467-9280.2005.01614.x.

Whetstine, Leslie, Stephen Streat, Mike Darwin, and David Crippen. 2005. Pro/con Ethics Debate: When Is Dead Really Dead? *Critical Care* 9 (6): 538–542. https://doi.org/10.1186/cc3894.

White, R.J., M.S. Albin, and J. Verdura. 1963. Isolation of the Monkey Brain. In Vitro Preparation and Maintenance. *Science* 141 (3585): 1060–1061.

White, R.J., M.S. Albin, and J. Verdura. 1964. Preservation of Viability in the Isolated Monkey Brain Utilizing a Mechanical Extracorporeal Circulation. *Nature* 202 (June): 1082–1083.

White, R.J., M.S. Albin, J. Verdura, Y. Takaoka, L.C. Massopust, L.R. Wolin, G.E. Locke, N. Taslitz, and D. Yashon. 1996. The Isolation and Transplantation of the Brain. An Historical Perspective Emphasizing the Surgical Solutions to the Design of These Classical Models. *Neurological Research* 18 (3): 194–203.

White, R.K., M.S. Albin, G.E. Locke, and E. Davidson. 1965. Brain Transplantation: Prolonged Survival of Brain after Carotid-jugular Interposition. *Science* 150 (3697): 779–781.

White, Robert M. 2006. The Tuskegee Study of Untreated Syphilis Revisited. *The Lancet Infectious Diseases* 6 (2): 62–63. https://doi.org/10.1016/S1473-3099(06)70357-7.

White, Robert M. 2006. Effects of Untreated Syphilis in the Negro Male, 1932 to 1972: A Closure Comes to the Tuskegee Study, 2004. *Urology* 67 (3): 654. https://doi.org/10.1016/j.urology.2005.08.025.

White, T.I. 2007. *In Defense of Dolphins: The New Moral Frontier*. Oxford, UK: Blackwell.

Whitehouse, Peter. 2011. Empowering Whom? Neuroethics at Its Limits. *The Lancet* 377 (9764): 468. https://doi.org/10.1016/S0140-6736(11)60161-X.

Wijdicks, E.F. 2001. The Diagnosis of Brain Death. *The New England Journal of Medicine* 344 (16): 1215–1221. https://doi.org/10.1056/NEJM200104193441606.

Wijdicks, Eelco F.M. 2002. Brain Death Worldwide: Accepted Fact but No Global Consensus in Diagnostic Criteria. *Neurology* 58 (1): 20–25. https://doi.org/10.1212/WNL.58.1.20.

Williams, B. 1985. *Ethics and the Limits of Philosophy*. Cambridge, MA: Harvard University Press.

Williams, B. 1993. *Shame and Necessity*. Oakland, CA: University of California Press.

Wood, Allen. 2005. Ethics and Embryonic Stem Cell Research. *Stem Cell Reviews* 1 (4): 317–324. https://doi.org/10.1385/SCR:1:4:317.

Wood, Mary F., Fredrick N. Nguyen, Michael S. Okun, Ramone L. Rodriguez, Kelly D. Foote, and Hubert H. Fernandez. 2010. The Effect of Deep Brain Stimulation Surgery on Repetitive Behavior in Parkinson Patients: A Case Series. *Neurocase: Case Studies in Neuropsychology, Neuropsychiatry, and Behavioural Neurology* 16 (1): 31–36. https://doi.org/10.1080/13554790903193190.

Wooldridge, D.E. 1968. *The Physical Basis of Intelligent Life*. NY: McGraw Hill.

World Medical Association. 2002. World Medical Association Declaration of Helsinki: Ethical Principles for Medical Research Involving Human Subjects. *Journal of Postgraduate Medicine* 48 (3): 206–208.

Wu, Fan, and Cheemeng Tan. 2014. The Engineering of Artificial Cellular Nanosystems Using Synthetic Biology Approaches. *Wiley Interdisciplinary Reviews. Nanomedicine and Nanobiotechnology* 6 (4): 369–383. https://doi.org/10.1002/wnan.1265.

Yesavage, J.A., M.S. Mumenthaler, J.L. Taylor, L. Friedman, R. O'Hara, J. Sheikh, J. Tinklenberg, and P.J. Whitehouse. 2002. Donepezil and Flight Simulator Performance: Effects on Retention of Complex Skills. *Neurology* 59 (1): 123–125.

Zehr, E. Paul. 2015. Future Think: Cautiously Optimistic About Brain Augmentation Using Tissue Engineering and Machine Interface. *Frontiers in Systems Neuroscience* 9 (May): 72. https://doi.org/10.3389/fnsys.2015.00072.

Zeki, S. 2007. The Neurobiology of Love. *FEBS Letters* 581 (14): 2575–2579. https://doi.org/10.1016/j.febslet.2007.03.094.

Zhao, Yong, Hui Jiang, Xin-Wei Liu, Liang-Bi Xiang, Da-Peng Zhou, and Jian-Ting Chen. 2015. MiR-124 Promotes Bone Marrow Mesenchymal Stem Cells Differentiation into Neurogenic Cells for Accelerating Recovery in the Spinal Cord Injury. *Tissue & Cell* 47 (2): 140–146. https://doi.org/10.1016/j.tice.2015.01.007.

Zubieta, Jon-Kar, and Christian S. Stohler. 2009. Neurobiological Mechanisms of Placebo Responses. *Annals of the New York Academy of Sciences* 1156 (March): 198–210. https://doi.org/10.1111/j.1749-6632.2009.04424.x.

Index

D. J. Doyle, *What Does it Mean to be Human? Life, Death, Personhood and the Transhumanist Movement*, Anticipation Science3,
https://doi.org/10.1007/978-3-319-94950-5

MIX
Papier aus verantwortungsvollen Quellen
Paper from responsible sources
FSC® C105338

If you have any concerns about our products,
you can contact us on
ProductSafety@springernature.com

In case Publisher is established outside the EU,
the EU authorized representative is:
Springer Nature Customer Service Center GmbH
Europaplatz 3, 69115 Heidelberg, Germany

Printed by Libri Plureos GmbH
in Hamburg, Germany